PRAISE FOR RAJU BHUPATIRAJU

"In my first conversation it was clear that Raju's approach is unique and refreshing. His team churned out large deals of the latest products every quarter even when the more advanced countries were struggling to match. He has a rare combination of high energy, passion, solution selling capabilities, and being an advisor to customers to consistently deliver outstanding results. It worked equally well in getting complex internal approvals also. I enjoyed many sessions with him hearing his strategy and approach."

—**Andrew Lim**

Managing Director, Government and Large Enterprise, Singtel Enterprise Business

"Raju took over a newly formed territory and overachieved within the 1st year. His approach of selling an end-to-end Oracle footprint through effective partnerships in the market has delivered results and growth in market share. He is also a straight shooter which I respect."

—**Lee Thompson**

Managing Director, Australia & New Zealand, Nutanix

MAGICAL SELLING

See with clarity.
Adapt to change.
Achieve goals predictably.
And thrive while others chase the "wait till things are back to normal" illusion.
That's Magical Selling.

MAGICAL SELLING

Engineering Enterprise Sales Success on Any Team, in Any Industry, and in Any Economy

RAJU BHUPATIRAJU

Power of Disruptive Solutions Pte Ltd
Singapore

This book is for informational purposes only. It is not intended to provide professional advice or to address all circumstances that might arise. Individuals and entities using this document are encouraged to consult their own counsel. The author and publisher specifically disclaim any and all liability arising directly or indirectly from the use of any information contained in this book.

The conversations in the book are based on the author's recollections, though they are not intended to represent word-for-word transcripts. Rather, the author has retold them in a way that communicates the meaning of what was said. In the author's humble opinion, the essence of the dialogue is accurate in all instances.

Power of Disruptive Solutions Pte Ltd,
30 Cecil Street, #21-08 Prudential Tower
Singapore 049712
+1 (760) 284-4477
www.MagicalSelling.com
Send feedback to Raju@MagicalSelling.com

Publisher's Cataloging-In-Publication Data

(Prepared by The Donohue Group, Inc.)

Names: Bhupatiraju, Raju, author.

Title: Magical selling : engineering enterprise sales success on any team, in any industry, and in any economy / Raju Bhupatiraju.

Description: Singapore : Power of Disruptive Solutions Pte Ltd, [2020] | Includes bibliographical references.

Identifiers: ISBN 9781735854601 (paperback) | ISBN 9781735854618 (hardback) | ISBN 9781735854625 (ebook)

Subjects: LCSH: Selling. | Sales management. | Sales personnel. | Success in business.

Classification: LCC HF5438.25 .B48 2020 (print) | LCC HF5438.25 (ebook | DDC 658.85--dc23

To family, friends, colleagues, competition,
customers who shaped my experiences,
and to everyone who aspires to improve
their selling skills.

GO BEYOND THE BOOK.

Ask Raju to help your team win enterprise sales deals like magic at www.MagicalSelling.com.

CONTENTS

Foreword

Over the last twenty-five years, I have worked with thousands of sales reps and sales leaders in my career as a solutions architect, in presales management, and in regional business development. My job was to help support sales teams create technology solutions for customers.

My engagement with Raju Bhupatiraju happened in the most unlikely circumstances and in the most unplanned manner. During the 2008 global financial crisis, Oracle created a new territory called South Asia Growth Economies that was assigned to Raju. These countries didn't have offices, and we did not have high expectations from the region.

Early in 2009, I got a call from Raju, who was in a remote country. He told me that a large telecom customer was interested in buying a brand new product. This customer had not bought from us for at least three years. This quick call for help ended up as a six-hour marathon call with more and more people added as the call progressed. Eventually, Raju closed the deal. It was one of the region's largest deals coming from an unexpected account.

This was the beginning of effective, high-energy account management that leveraged different support and technical teams to generate a "sweep the table," with win-wins over four year of Raju's sales leadership. During that time frame, we grew more than 8X in sales revenue, and actual sales even surpassed many large countries in the region.

Raju's approach was always thought of as unique. I'll let you read through the details in this book. He followed a method that energized all the support functions, provided clarity on what's required to win no matter what odds you start with, and propelled everyone's career on that team.

Anyone following Raju's methods will be able to bring their A game to every deal effortlessly. I recommend this book to anyone in customer-facing roles to bring out the best in every sales team.

—**Joshua Chua**

Pillar Lead, Enterprise Data Management, JAPAC at Oracle

CHAPTER 1

Become the Most Successful Salesperson on Your Team

I never intended to work in sales.

When I was in college, nobody even knew that "sales" was a formal career. If you told someone you worked in sales, they assumed you went door to door selling vacuum cleaners. In those days, we considered finance, medicine, and engineering the ideal job fields.

It's probably no surprise that I, like many others of Indian descent, earned an engineering degree, but the general perception that all Indians are techies has never been true for me. I don't have the technical savvy that most people from India seem to have. So, what do you do with an engineering student who doesn't want to be an engineer? You send him to graduate school for an MBA!

That's where my sales story actually begins. In the summer of 1997, I was working in market research for a denim manufacturer on a short-term summer project while I worked on my MBA. I spent two months learning about the product, the pricing, and the positioning. I hoped I'd soon get an offer to work full time in their advertising or marketing department. At the end of the two-month contract, my wish came true. The company's vice president gave me and another student worker preplacement offers, meaning he wanted to hire us as full-time employees after we graduated. The prospect of having guaranteed employment made me extremely happy. Of course, if something better came along, I'd take it. At school, I was helping to organize on-campus career fairs by designing and distributing brochures, and arranging for companies to meet with students. I had everything I wanted: a good job waiting for me after graduation and the first pick at any better opportunities that might come along.

If you remember the headlines from 1997, you already know what happened next: the Asian financial crisis. All of a sudden, companies that had confirmed participation in campus career fairs cancelled. Our professors assured us that the currency declines, falling imports, and stock market collapse would all self-correct in no time. That's why the news I got in November came as a shock. The denim manufacturer was being acquired by another company.

What about my job? I hadn't bothered interviewing with any other companies because I thought that job was a sure thing. I called the VP, but he couldn't confirm anything. He didn't know what to tell me, nor did anyone else at the company. It felt like my future was crumbling. I got a few interviews with other companies over the next month, but nothing worked out. With no cushion, no cash savings, and no sign of economic recovery, I was getting desperate. I refused to ask my parents for help after graduation, even after it became clear that my job at the denim manufacturer was gone.

As I scoured the jobs section of the local newspaper, I saw an advertisement with one word in big, bold letters: "Grow." Xerox, the photocopier company, was looking for sales trainees. It was a temporary job, and it didn't pay well, but with less than 6 months to my graduation what other options did I have? At least Xerox was a well-known brand. If they hired me, I'd at least be able to support myself. If I excelled, I could earn extra commissions. So, I swallowed my pride and applied for a sales trainee position.

I got the job. It turned out these Xerox salespeople were real go-getters, my boss especially. For six years in a row, his team in India was the Xerox national sales champion. That's saying something for a big company with a lot of competition. Once might have been luck. Two or three times would have been impressive. But six in a row? Maybe I could learn something here.

During my sales training, I learned all about photocopiers and also about some guy called Daryl, who was my boss. He was a legend there. The trainer and all the full-time salespeople told stories about Daryl. He turned nos into yeses. He closed deals in the middle of the night. He went into companies that already had photocopiers from competitors and walked out with sales to replace them all. In short, Daryl was a shrewd salesman and a wise sales manager, a powerful combination.

As luck would have it, I was assigned to Daryl's team to continue my sales training. We worked five and a half days a week. Saturdays were for team meetings and learning. During these Saturday meetings, Daryl analyzed each and every team win and loss. I liked his approach because everyone got to learn from everyone else's achievements and mistakes.

Several days into training, I was given my own sales territory, officially making me a full-time sales guy responsible for generating revenue. Daryl took me aside.

"I know you have a better education than a lot of these guys," he said. "But you're still going to make mistakes. I *expect* you to make mistakes. Your job is to recognize any mistakes you made. If you identify any mistakes, keep me informed so I can help you clean them up. I can't help you if I don't know they exist."

At that moment, any fear I had about my new job disappeared. I could literally feel my confidence grow knowing he wasn't going to judge me. I could speak freely without worrying about getting fired.

Working with Daryl let me see his magic with my own eyes. Deals I never expected to close, I did. I engaged with the CFO of a financial consulting firm looking to upgrade her office's photocopiers. I offered our midrange machine, but a Japanese competitor had pitched her a better deal—double the specifications for half the cost. She wasn't happy with my price even after I offered a discount.

"Why would I be interested in Xerox if someone else can sell me a better photocopier at a lower price?" she asked.

I wondered if I should give up. I wished the CFO a good day and walked out. Then I went straight to Daryl and told him everything.

"I'm not authorized to give her any more discounts," I said.

"Why don't you schedule an appointment for her to meet with me?"

"I tried that. She didn't want to talk to you unless you offered her a cheaper price."

"Continue with your other sales calls," Daryl said. "I'll make the appointment myself."

That was at ten in the morning. Daryl paged me later, saying he had an appointment with the CFO at four that afternoon and that I should be there. My scooter got me there fifteen minutes early, but Daryl was already waiting in the lobby.

"Why are you so late?" he asked.

"I'm not late! I'm fifteen minutes early."

"Whenever you're going into a negotiation, never rush yourself. You need to be able to think clearly."

"OK," I said. *Isn't fifteen minutes plenty of time?* I thought.

Daryl and I could see through the CFO's office windows. She was talking with someone. At exactly four o'clock, she stepped out to address us.

"Gentlemen, I'll be right with you. I'm still finishing another meeting."

After she returned to her office, Daryl asked me, "What do you think of that?"

"I don't know. What is there to read. . ."

"Pay attention. That's a very respectful way to treat people," he said. "She had the courtesy to pause her meeting, come out here, and ask us to wait until her meeting is over. This sale is not lost. It's wide open."

A few minutes later, the CFO invited us into her office. Right away it seemed as if Daryl and the CFO knew each other. I knew this was their first-ever meeting, yet they hit it off like old friends. They talked about everything but the sale. I didn't understand. I was glad to see them laughing and joking, and even happier to be a part of it, but I had no idea where the conversation was going.

Nearly two hours into their banter, they had occasionally touched upon the sale but did not get specific about pricing. Then the CFO finally touched on the product.

"I'd love to do business with you," she said, "but I've got a counteroffer." She then said she was willing to pay a price that was ten thousand rupees more than what she had told me previously. Was this a slip of the tongue? Had she forgotten the price I gave her?

"I understand," Daryl replied. "I just want you to know that no other machine can beat the functionality of ours. And you have the reliability of the trusted Xerox brand backing it up. What's going to happen four months from now when that other machine breaks down? I know they're not offering the same level of warranty and support that we are."

Just then, the CFO's secretary interrupted us. A visitor from their Mumbai office arrived and the CFO excused herself.

Daryl turned to me. "What do you think is happening with this account?" he asked.

"Well, you guys are clearly getting along well, but I'm surprised you're not talking about price. She told me not to waste her time if we couldn't reduce the price. So. . . I guess I don't think we have a chance of closing."

Daryl smiled. "We're closing this account tonight before we leave."

"How?"

"She likes us," he said. "What have we been doing for the past two hours? We've been getting to understand her needs better than the Japanese company does. And she knows it."

"I must have missed how to do that in training."

"You can't train this," Daryl said. "It isn't theory. There are no 'right' questions to ask to make someone like you. In fact, questions can be counterproductive. We haven't talked about pricing because pricing was never the problem."

The secretary came in to fetch a folder from the CFO's desk. More small talk. Within a minute, Daryl had the secretary laughing.

"And I'm sorry for the interruption," she told us. "She shouldn't be much longer."

The CFO stepped back a few moments later but not before Daryl had struck up a conversation with the secretary.

"Miss, the CFO showed me my value. It's ten thousand rupees," Daryl said in front of everyone, referring to the CFO's counteroffer.

"I didn't mean Daryl's value was ten thousand rupees," the CFO said.

Then I recognized Daryl's strategy. He wanted her to move the goalposts from the previous number I had quoted her to a higher number. In a way she just acknowledged that Daryl was more valuable than the additional ten thousand rupees she offered. He was priming her to acknowledge that the value of our photocopiers was higher than what we'd already offered her.

"That's a good deal," Daryl said. "And that means the deal Raju offered you is fantastic. We can give you the discounted price *and* include three reams of paper at no additional cost. They're yours when you place the purchase order. Would you like to do that?"

"I do like your machines better," the CFO said. "And your competitor can't beat your customer service. Let's do it. But it's late. I'll have my secretary print the PO tomorrow."

"Since we're all here, I'd rather take care of it now," Daryl said. He opened up his briefcase and wrote down a proposal on the letterhead. "Can you sign off on this so we can leave with the order tonight?"

After a bit of resistance, "Oh. Sure," she said.

Daryl was obviously pressuring her to seal the deal but in a smooth manner. It takes a lot of experience to think on your feet and pull that off. The CFO gave Daryl's written proposal to her secretary to type up. While we waited, Daryl asked the CFO about her leadership style.

"You're the CFO of a successful financial services company," he said. "I'm sure you have some tips I could use in my own line of work."

The CFO answered his questions but not with anything memorable. A few minutes later she signed the proposal, and we left.

"Congratulations," Daryl said once we were outside. He handed me the signed purchase order and asked me to go have a beer.

"I can't take credit for this deal," I said. "You did all the work. You predicted exactly what was going to happen, and it did. And you didn't drop below my price.

When I walked into the office the next morning, Daryl announced to everyone that they should congratulate me for closing that deal.

"It wasn't really me," I muttered. "It's Daryl who should be congratulated."

"Look, Raju," Daryl said. "Your job is to start the work, and my job is to help you finish it."

I didn't know it at that time, but I would practice Daryl's approach for the rest of my career. He started engaging with the account exactly where I had hit my limit. Until then he would just monitor the progress. Now, I consciously try to identify what I need to do to complement each team member, each partner's team member, or anyone else I am working with. I try to find out what they can do and help them with the parts where they need help. People have different strengths in a sales cycle, such as, prospecting, planning, creating solutions, negotiating, closing, and so forth. I make it my bread and butter to fill in the gaps where they are required for every deal. That's why I've never had a problem with working, coaching, or managing people with a wide range of strengths, weaknesses and experiences. As long as they learn from each engagement, my involvement becomes less and less, leaving me to work on other harder deals.

A few days after that deal closed, Patrick, a senior sales guy on Daryl's team, asked me to go with him to meet with an educational engineering institute. The dean was a former military colonel.

We got to their office only to realize the Purchase Order was issued for some other vendor. The clerk made a mistake of giving it to us because those days all photocopiers were called Xerox machines.

Patrick looked at the PO, said something about not liking the payment terms, and said he can't collect the PO without internal approval. He dropped the PO onto the administrator's desk and walked out and informed Daryl.

"Tell me what happened, from the beginning" Daryl asked.

Patrick said. "He sent three professors to see our demo, but he attended the other vendor's presentation himself."

What followed was another eye opener for me. Daryl said to the colonel on the phone. "I thought that people in the military made decisions based on intelligence, not hearsay." Daryl connected the Military background to the gap in the evaluation process, in an attempt to get the 'apparently lost' deal back to the table.

I didn't hear what the Colonel said but Daryl gave his name again on the phone. The dean agreed to meet with him. Daryl met with the dean for a one-on-one meeting and closed the deal an hour and a half later.

Watching Daryl close deal after deal changed my life. I took his almost magical approach and made it my own. I used that trainee job at Xerox to launch a real career in sales. Those experiences with Daryl taught me that if you pay attention to the customer's signals, you can understand what they want, find the right angle, and succeed. I only worked as a graduate trainee for five months, but that time I spent learning from Daryl was far more valuable than my formal university education.

The training wheels were off, and the real journey began.

I finished my MBA a month later and my sales career took me all over the world. I got a job in computer-aided manufacturing software solutions, better known as CAD/CAM, at a reseller in western India who sold the software and services from US companies. I moved to manage government accounts in Hyderabad. I then moved to Chennai to sell similar products for a different company. Later, I sold cloud-based e-learning platforms. Back then it was called Hosting Services. Not cloud. There I started up operations for new businesses in Hyderabad and in New Delhi.

Four years and three months after graduating from business school, I got a job at Oracle in Bangalore. That was when my career took shape. In my fourth job in about four years, I started selling solutions in some of the toughest territories in India, where budgets are a challenge, sales cycles are long, and manufacturing industry sales are hard. Oracle gave me a newly formed territory called "general business." It had a lot of small- and medium-size businesses as well as any business that was not among the major verticals. Oracle's primary business was telecommunications, media, utilities, financial services, insurance, manufacturing, and the public sector. I noticed an opportunity to offer other services, especially for the outsourcing arms of large corporations in the United States. Three years later, I moved to Singapore to work at Oracle's regional headquarters. Eventually I managed a thirty-plus-person sales team with revenue targets up to $56 million.

I didn't realize it back then, but there was a pattern to what I was following. At every company I worked for, I either started up new territories for existing companies or I "fixed" struggling territories. I became a specialist at turning things around. I worked with transactional sales, solution-based selling, value selling and transformation deals too. How do you highlight the compelling value propositions? How do you justify your higher costs? How do you justify implementation cycles? I found the answers to these gaps in large transformation projects—the largest of them being worth $32 million.

In a way, I had "made it." My results were some of the best, and everyone knew it. At that time, I couldn't articulate what I was doing differently from everyone else. Everybody knew my approach was different and effective, and they'd comment on what they perceived, but when junior salespeople asked me for advice, I didn't know how to articulate the difference in my approach in a short and crisp way. I was inherently uncomfortable trying to describe my methods. What's the point of thinking that my approach was different when i can't articulate it in simple terms for others to understand?

Maybe Daryl was right. Maybe they couldn't be taught. But I am the living proof that they could be learned. How could I bridge that gap?

I had the opportunity to build businesses in South East Asia, India and the Middle East. I did Asia Pacific sales and more recently a Global sales role. It involved selling to corporations of all sizes, solutions that were mature, Solutions that were cutting edge and even selling to a cluster of companies, not just one at a time. Selling to one enterprise is hard and time consuming. Imagine going through all the efforts of selling to three or four different enterprises simultaneously. Imagine hiring for all those roles, building partnerships, coaching teams, delivering sales quotas quarter after quarter.

All those varied roles helped me realize how to show others how to become the best salespeople they could be. There wasn't a magic script or template I could hand to the people on my team, but I *could* teach them the magic of selling. What had Daryl done during my first few months at Xerox? He led by example. Now it was my turn to do the same.

That brings me to the present day, to this book, and to why I decided to write it. At the time of this writing, there are over 50,000 sales self-help books available for purchase. Why write another? Because I have something to say that those other books haven't.

Even though I was given constant sales training - many theories, practices, sales templates, and word-for-word scripts—none of it prepared me for the real world. It's the same for the thousands of other salespeople I've interacted with over the years. They keep reading sales books and undergoing professional sales training, but they don't detect any measurable impact on their day-to-day life. Cold prospecting doesn't get easier. Leads don't object less. And their closing rate doesn't go up. Sales is still a 'rejection' business 9 out of 10 times. It's true that some sales training programs can help, but they only offer bits and pieces. And nothing out there fits every situation.

Sales people are branded into many different types. There are hunters, farmers, problem-solvers, relationship sellers, solo fighters, solution sellers, consultative sellers, value sellers, deal makers, transactional salespeople, and more. I'm sure you've also noticed the diversity of your potential customers too. When it comes to sales success, customers' different backgrounds, industries, and countries only add to the complexity of finding a "theory of everything." What is cutting edge in one country may be old news in another. One-size-fits-all sales strategies, tactics, and mindsets cannot fit the countless different situations you'll encounter over the course of your sales career. There is no such thing as the ideal salesperson, the perfect sales personality, or the sales superstar archetype, although employers always look for them. The reason I couldn't "teach" what I was doing before was because it's not *about* teaching. It's about leading by example. Daryl led by example to reveal that how you approach one prospect does not carry over to how you should approach another. My own experience has confirmed this and led me to coin the phrase, "Sweep the table." Why does my approach work whether I'm running a large company or selling new solutions or new markets? Whereas people get branded as certain types of seller, I was at home in every opportunity, most of which were wildly different. So-called "different cultures" were also thrown into the mix as I moved around the globe. The reality is I was following—or rather, engineering—a system that was based on '*First Principles*'. It was developed over time through all of my different experiences.

If you are new to systems-based thinking or the superiority of systems over goals, I recommend you read Scott Adams's book *How to Fail at Almost Everything and Still Win Big*. Over time, I realized I was developing a systems-over-goals approach to selling. As I see it, the goal, which is to close a deal, is the consequence of doing a few things properly during the sales cycle.

When leading sales teams, I have used the following phrases to describe systems-based thinking for sales success in different contexts:

1. **"Sweep the table."** Sweeping the table is a system that nudges sales reps to think about what else the customer is buying or could buy, how you can influence that to happen. Enterprise projects commonly involve multiple vendor products to work together.. To "sweep the table" is an expression from poker that refers to winning all the bets on the table. You don't need to know anything about gambling to understand how difficult this is.
2. **"That's the byproduct. What is the main product that creates the byproduct?"** Clarity on this question is like an instant eye opener or energizer for any company. Sales reps get caught up in the goal. Calling it the byproduct of what they do allows them to focus on the main product, which is the system.

3. **Fort and Moat:** This is for territory-related discussions with internal management to justify discounts or resource allocation and headcount approvals. Convincing management means providing clarity on who we're defending (**"Fort"**), and preventing competition from attacking us (**"Moat"**). In other words, it simplifies our ability to take the fight to the competition's doorstep. Winning a deal is the goal, but how we justify and get approval is the system.

What is the big difference that systems-based thinking leads to? When you only focus on goals like winning a deal, it's either a one or a zero. Until you get to the one, the win, you live in perennial stress. It's not done until it's done. For systems people, there is forward progress in each stage of the sales cycle. They tend to be more energized and happy during the sales cycle and think more clearly and more objectively. That helps with career longevity. Have you noticed that some sales reps seem to be energized more than the others? The reason is likely engineering. Their precise approach to selling makes it look like they are having fun. Look at the stressed sales reps, and you will know they are likely focusing on goals instead of the system that will get them there—either that or their system itself is not good enough.

Not many prime salespeople are stress resistant enough to thrive for fifteen years at Oracle. Systems approach based on first principles is the reason I did. Systems-based sellers work better under pressure. Looking at the final "goal" of a sale alone stirs up stress leading the salespeople to find it hard to go through rough forecast calls. Instead, if we focus on the system that is built from the first principles (not based on analogies), we're more relaxed as we make material progress every day.

Elon Musk once said, "Engineering is the closest thing to magic that exists in the world." This applies to selling as well. That's why I titled this book *Magical Selling*, which refers to the result of its lessons: an almost magical ease in closing sales. Behind the curtain is an engineered system based on first principles that can be easily applied in all enterprise selling scenarios. The four-part system outlined in this process is simple to understand. Focus on this, and within a few deals, I'm confident you will see your energy go up. . . like magic.

If it can't be taught and can only be learned by example, how can this book help you?

Since the inception of my career, I've searched for a sales method that makes sense to everyone, that allows anyone to look at the uniqueness of every opportunity to see what is needed to win the deal. I have found no such method in any book or training, but I've found a way to create such a method that *can* be learned. I only wish I'd created it sooner.

The key is in the title of this book: magic. Everyone has a preconceived notion of what magic is. For instance, I asked my daughter to tell me what she thinks about when I say the word "magic."

"Magic?" she repeated. "Like black magic? Magic pills? Or magic tricks? Like on a stage where the guy shows you a card and makes it disappear?"

I want to thank my daughter for a thorough answer because that is *not* what I mean by magic. Stage magic is about misdirection. What else might come to mind? Cheating. "Tricking" the audience. Manipulation. I will advise nothing like it in your selling efforts. Instead, I am using magic to describe the ease of the outcome. The way you close deals that no one expects of you will "look like magic." That is by design. Anything that is well engineered looks like magic. That is *magical selling*.

Just as every engineered system includes multiple elements that work together, *Magical Selling* consists of two parts. First is how to think (a *Thought Plan*) and then how to execute it (the *Execution Plan*).

The five aspects of the Thought Plan give you an indication of where to focus. It is important to note that these aspects of the Thought Plan need to be performed in tandem from the time of engagement. I think of this as a chain, and the weakest link decides the strength of the chain. Therefore, these aspects serve as an early warning system for course corrections, including go or no-go decisions. I have used this framework to describe what's going on in my territory. You can actually do it in one slide in a presentation to set the table. If someone asks you what your business plan is or even in an interview, try using this rubric. You will automatically look like a great presenter!

MAGIC Thought Plan (Ready!)

Let me introduce you to the all-weather sales system first:

- **M** is for **MAN**—Money, Authority, Need—and refers to people you need to cover. This can range from one person to dozens of people depending on the size of the organization and its average deal value. Instead of getting into designations that can be misleading, many times I find it effective to check if the person we are talking to is an M or an A or an N. Do they control the budget, are they the decision-maker, or are they the one with the need? We'll get granular on this topic soon.
- **A** is for **Alignment.** Every deal is unique. Look to align with the customer's People, Process, and Technology with your own resources. Alignment is a two-way street and applies to your company just as much as to theirs. It starts from the time you qualify a deal as real. Successful salespeople are good with internal and external negotiations, and alignment is where you start setting the table for these talks. If done correctly from the beginning, you can save big on time and effort. This helps you set expectations on both sides to help find common ground at a later date.
- **G** refers to **Grow** *your influence* in the opportunity. Your initial position in any sale may be negative, neutral, or positive. Actively discover that and plan to grow your influence. It's like a self-awareness check in a sales cycle that helps identify all that is required. Do you need a partner? Who on your team is best suited for a task? Do you have champions that you can leverage in the customer's organization? And so on.
- **I** is for **Inspection**, inspection, inspection. You can expect what you can inspect. Inspect all steps and activities for your deal daily. This is the discipline and cadence you need to bring the sale home. Most people follow a weekly cadence on forecast calls and so on, but I recommend that you inspect the progress every day. Once you start doing it, you will realize it takes much less time to man-

age and gives you greater control over course corrections. This applies to territory, accounts, and/or teams.

- **C** is for **Context** is the only thing that matters. Curiosity is the catalyst to get the context and generally a good indicator of how well you listen. When you are curious to find out the customer's context for their buying decision, you will automatically listen more and pay more attention. I call it the "herd mentality." It's the customer who defines who your competition is, so I'm eager to find out what we are up against every time. I have never found a deal boring because every bit of context is different and is a fascinating aspect of a sale. Understanding context and tracking how it evolves in a sales cycle will make you a wise decision maker and leader.

Each letter in MAGIC can be used as a separate lens through which you can view the remaining four letters. M is a lens to cross-reference A, G, I, and C and so on.

MAGIC Execution Plan (Aim! Fire!)

Or "Fire and then aim" in some cases where it is applicable. This is a process of bringing together four important aspects of a deal that give you the desired results. The MAGIC Execution Plan has to do with sales discipline, teamwork, real-time responses, and thinking on one's feet. You can expect to be more successful in this execution plan if you've already done the footwork of the Thought Plan.

Unlike the Thought Plan, the Execution Plan is not an acronym because MAGIC maps perfectly onto the actions you'll be taking as you execute. It's theory versus action. Any book can offer sales tips but this one is founded on a time-tested first principle: MAGIC. With the lack of mnemonic device explained, let's look at the Execution Plan:

1. **ACT 1: Understand your customers:** How do you "read" a customer so you can build instant rapport? If you're already good at this, this section will boost the success you've already achieved. If you're new to sales or you're struggling with understanding your customers, this section will shape your life.

2. **ACT 2: Validate that customers know you understand them *better than your competitors do*:** The key part is "*better than your competitors do*" and very few sales people do it. Simply knowing a customer's needs is insufficient in some cases. To make your customer give you the "first right to refuse", this process is critical. This section will show you how.
3. **ACT 3: Build partnerships:** Partnerships include both internal and external relationships. Enterprise sales involves a community of your company employees engaging with a community of your client's employees. If you work for a corporation or sell to them, these partnerships will involve people of many languages, countries, and histories. Going at it alone is typically not scalable and in my experience doesn't maximize your revenue. In the partnership section, you'll learn how to do just that.
4. **ACT 4: Win with a plan:** Identifying your odds of winning an opportunity upfront and working to improve those odds as you go along is a skill. In this section you will learn how to measure your odds and how to plan a win.

The real magic is taking these four strong solo acts and making them all work together

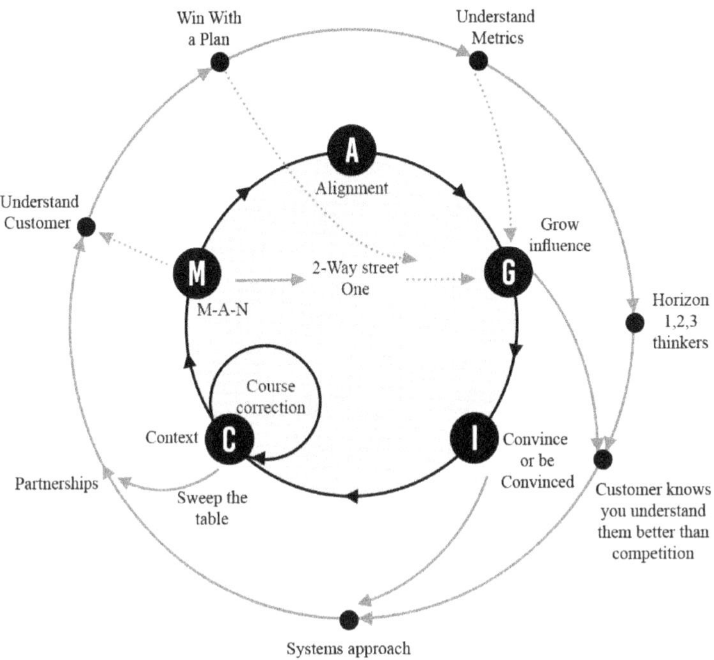

Figure 1 The Magical Results Flywheel.

All high-performing salespeople incorporate these four practices into their career. Engineering sales success is not magic, although it may look that way to the uninformed. Daryl's salesmanship seemed magical to me until I made it my own and later deconstructed it. In reality, it's simply a process that anyone can follow with any customer. You can upsell any account. You can "sweep the table." You can sell stuff nobody else in your company has sold and sell it in new territories. You can even succeed across countries, cultures, and languages.

A few words about intercultural sales. The culture in every country is different. Most of my career I have spent in the Asia Pacific region. Over the last couple of years, though, I've also made deals in the Middle East and in UK markets. In principle, people are different in different geographies. Thai customers differ from Japanese customers, and Australians differ from everyone else in the British Commonwealth. Still, everybody running companies in these countries has specific needs at any point in time. As long as your offer is relevant, what you read in this book will work anywhere with anyone.

Once you internalize this as your one technique, nothing else will be required beyond what you'll learn in this book to succeed in sales. I say this with certainty—I'm no Brad Pitt, so people don't take one look at my face and buy whatever I'm selling. I have to put in the work, but it's not guesswork. It's a process anyone can practice right now.

Magical Selling will also teach you how to coach others, which means you'll become a manager sooner. But first, I have much to unteach you. Yes, you read that right. In sales, just like everywhere else, you have to unlearn the things that really don't work. Now, they may work in certain contexts. Never say never. Be open, be flexible, and be malleable. Don't take a firm position when it comes to selling. Why? If you're rigid, you can break. If you're flexible, you can't. Customers like communicating with people. It's never about the product alone. If I don't buy a Honda, I'll buy a Toyota. They're both good cars. I might say it's budget. I might say it's the features. But in the end, what am I doing with the car? If the guy selling me my car understands that I just need to drop the kids off at school, he won't offer me a Formula One supercar. But when a racecar driver shows up, he'll offer him the fastest car he has because that's what the driver wants.

As you unlearn these lessons, you're going to stop making unconscious mistakes. A lot of people are screwing themselves, and they're clueless to the fact that they are. They think they're being smart, and they wonder why their career isn't taking off, why their bosses aren't giving them a raise, or why they can't get a new job.

As you gathered from the title and from some of my previous explanations, this book is written with enterprise sales in mind. This means selling to a business entity rather than to an individual. Selling to an entity or enterprise includes all sales situations where the customer is the name of a company. This book will benefit anyone that is selling to an enterprise, irrespective of whether it's in a small or large company.

There are push sales and pull sales. With pull sales, you don't spend too much time talking to people. If Cadbury wants to sell chocolate, they don't talk each individual into eating chocolate. They advertise and focus on retail. Push sales, on the other hand, take several months of effort. Or even years.

As you can see, sales is a topic that cannot be covered without nuance. That's why this book separates marketing and sales. They're both addressing a go-to-market strategy. But marketing is the early part of the cycle and has a spending quota while sales has an earning quota. Some companies are big on sales, and some are big on marketing.

So, I'll leave marketing books to marketing authors. That said, I can promise you great sales outcomes from my book. Maybe you're reading this book to identify what you are doing well and not doing well. Whatever the case, by the time you're done reading, you will be able to use what you've learned in your very next interaction with a prospect. Implement the advice in this book, and you can expect to:

- Boost your sales commissions even in a downturn;
- Make informed decisions on where to spend time for maximum results;
- Know who you should keep in the pipeline and who you shouldn't;
- Speak articulately to customers and management;
- Become well known in your business and industry;
- Manage your sales team (management included);
- Have higher levels of energy instead of stress;
- Sell to customers from other countries and cultures;
- Design a sales acceleration formula that motivates your team;
- Predict which deals are at risk;
- Troubleshoot quickly and accurately;
- Grow your business and realize your career potential.

You might be thinking, *Well, this might work for me, but. . .* I call these the "yeah, buts." And there's merit to your concerns. You haven't seen the process for yourself yet. When somebody says to me, "This territory doesn't work," "This account doesn't work," "My problems are the worst," or "I'm having the toughest time," I firmly believe that it's all contextual. No two accounts are the same, even if you're selling them the same thing. If you ever find yourself saying, "Raju's book doesn't work," you have my permission to email me directly with any questions or doubts. I'm completely serious because I take you seriously. Reach me at Raju@MagicalSelling.com. I read every email.

I became a sales success because I learned by example. Now I'm going to teach by example. I'll share with you what I've learned through the details of real deals I've won, lost, and won back. I'll let you learn from them just as I did, as though they were your own experiences. No theories, no hypotheticals, no assumptions. Just the facts, and they work quickly.

Why will this new system start giving you results immediately? My advice is not a gamble on some radical new methodology. I have adapted the very best proven techniques and seamlessly integrated them into one seamless system that any salesperson can understand and implement in their career. In other words Magical Selling is based on First Principles instead of analogies

Now, it's time to understand where customers are coming from. The first section of this book is all about understanding customers, the first act to magical selling. I sincerely hope this book helps you distinguish yourself from the mediocre sales reps. Each one of the following four parts in this book are like solo acts, and it's up to you to make all of them work together. Let's begin.

ACT 1

"You don't really understand a person unless you know what they really want"

CHAPTER 2

Understand Your Customers Better Than Your Competitors Do

"You may be selling a drill bit, but the customer is buying a hole."

Enterprise selling in many ways is similar to team sports. Not like individual sports like sprinting or throwing a javelin or shooting. It's like those sports where an opposing team is actively trying to prevent you from moving forward. In sports like soccer, basketball, and cricket, your ability to win depends on who you are competing with. You have to figure out how to outsmart your competition in every deal. Your competitor is not your enemy. They simply keep you on your toes and, in a way, help you improve your quality. So, an inherent part of your effort in enterprise selling is always to outsmart your competition. One surefire way (not the only way, of course) to get that is to understand your customers better than your competitors do.

Understanding your customers is mandatory unless you are an order taker at a fast food counter. The overarching theme I realized in engaging with over a thousand customers in my career is that customers make a decision to buy a certain product long before the formal decision is made. From the time they decide, barring a few exceptions where the preferred solution vendor makes tactical errors or is complacent, the rest of the work is to put the data together to justify their reasons for buying. In other words the sale is for the preferred vendor to lose. The trick is to be the preferred vendor.

How do you achieve that? It's a combination of your credibility and proof of performance. Part of that credibility comes from your company and/or product; the rest is how effectively you go about understanding the customers' needs.

It's easy to start. The first step is to know your solution (*and* the problems it solves) very well—that is, what kind of drill bits you sell, who buys them, what kind of holes they drill and why, how do they drill holes today, what other methods there are to drill holes, and how efficient those methods are. What are the complementary products or services customers typically need when drilling the holes they need? If the customer uses your drill bit, what quantifiable benefit do they get? This might be measured in terms of energy consumption, longer life, time saved, and so on. Once you know this well, it gives you the ability to ask probing questions without having to memorize questions to blurt out in front of the customer. It's easy to connect with the customer that way. Often I've seen sales reps take a totally counterproductive approach. Everyone tells you to ask the right questions—and there's a big list of right questions. This is very counterproductive. **The prerequisite to asking the right questions is to fully understand your own solution and how it is used by customers.** It's futile to simply memorize a list of questions only to wonder why it's not working and what's wrong with you. Nothing is wrong with you. **You need a screwdriver, but you are using a hammer.** If this is the case, it's time to get someone to educate you really well on the product and the nature of the problem it solves. Understanding drillbit makes you a product specialist. Understanding everything about holes makes you a subject matter expert. Understanding both makes you win deals like magic.

Then comes context. Having an accurate context of your prospect's situation is critical to the outcome you want—a sale. If you were to look back at any deal from your career's "highlight reel," you would realize that you understood the customer better than any competitor did. The bigger and more complex the deal is, the more your team has to spend time to understand the customer's needs.

Those "highlight reel" sales probably taught you that finding out what makes a customer tick is not as straightforward as asking a few questions, yet that is precisely how most salespeople are trained to "understand" the customer, myself included. Early on in my career, prospects picked up on those scripted conversations, and they didn't like them.

Start looking at typical customer buying cycles for context: There are six logical stages

Unawareness → Awareness → Interest/Education → Evaluation → Justification → Purchase

If you're in a transactional business in which customers are highly educated, your engagement starts in the second half of this buying cycle. If a customer comes to you for additional licenses of what they already purchased, it's a transactional sale, even if your product itself is a complex one. These are short sales cycles. In today's world of forced virtual sales, it is safe to assume that probably 60 percent of customers get educated first, and companies need to be a part of that forum to qualify for the final stages of the purchase.

The other extreme is when you are in the business of creating a new category. Sales cycles are longer, and you start with creating or uncovering a need. First step is to establish where in the buying cycle your prospect is when they first speak to you.

Exceptions are possible in both extremes. A market leader entering a new segment or industry, may have to engage earlier in the awareness space. Similarly, for a totally new segment, a prospective customer might approach you after doing research, and that could be a quick sales cycle too. The stage at which you engage in the customer's buying cycle determines how you allocate your time and resources to reach your revenue goals.

In any large company, you'll find one sales rep may be selling into an industry vertical or a territory that may have a lot of transactions, and yet within the same team another rep may have sales cycles that are much longer in a new industry or territory. One rule of thumb I tried to follow all my career is to spend 30 percent of my time and my team's time closing current quarter deals (to achieve the quota, of course) and 70 percent of the time and effort on creating or progressing deals for future quarters. I hit this mark on a few occasions, but mostly it was a fifty-fifty split. If you notice yourself spending 90 percent of your time working on current quarter deals, you better realize that it may lead to burnout. It takes two to three quarters to move from there into a fifty-fifty or even a seventy-thirty split.

By focusing on deal qualification and going deep, your chances of winning the deal go up. If you are good at deal qualification, there is no need for a 4x pipeline like many companies practice. Chasing unqualified deals is a wild goose chase. Having a 4x or a 5x pipeline for 1x revenue generation is a sign that you do not understand the uniqueness of each opportunity.

Win probability goes up only with your full involvement. In my experience, sales reps and managers who regularly make their quota clubs or president's clubs usually have a 2x quality pipeline, along with a deep understanding of their customers. If you stop wasting time on unqualified leads, and instead invest your sales efforts in getting to know fewer qualified prospects, a 50% conversion is not uncommon. If deals are unqualified, update your CRM accurately so no one gets a wrong impression of your pipeline. The risk of losing deals, timeline risks, and other exceptions need to be accounted for in the CRM appropriately.

"Keep talking to customers" is my motto. By talking to customers, I really mean *listening* to them for context, with curiosity. Pay attention to three primary areas to listen to: **people**, **process**, and **technology**. That enables you to find the **measurable benefit** you can offer them.

All customer's projects are usually found in one of four buckets:

- A revenue-generating initiative
- A cost-cutting initiative
- New product or service development
- Customer success measurement

What happens if your customer can achieve a 1% improvement in their lead conversion? Or 1% reduction in fraudulent insurance claims? Automatic transmission cars resulted in reducing the insurance premiums by 50% because it reduced the number of accidents. Automation has its benefits and you have to find a way to measure those outcomes.

Every department involved in a purchase decision has a measurable success standard for a project. Even if it's not enforced by the company, it's your job to find out what those measures look like. A business user's metric may be different from that of the analyst or the developer.

Think of your account as an iceberg. Above the water are the business aspects of the customer's people, process, and technology. Below the water line is the vast majority of complex aspects such as legal issues, Information security, privacy, skills and technology preferences. Make an attempt to know all of them.

What if the customer's precise numbers aren't known? If your prospect says they want to improve their marketing, try to find out what their marketing results are now and what they want them to be in the future. Now, most prospects are not ready to share these details with you in the first meeting. They may not even know the metrics you ask them. Approach their current situation with tact and work toward getting directional accuracy. Every salesperson should either develop this skill or leverage someone who already has it. If you sense that customers are unprepared, try discussing estimates, such as a 20 percent increased return on ad spending. Or educate them how another customer may have measured it. This brings you and your prospect onto the same page. Customers will sense your commitment to helping them figure out your offer's measurable benefit, and as a result they'll feel motivated to continue your engagement.

Does this scenario seem too easy to be true? I'm not suggesting that selling is easy, but if you consciously work to understand your customer deeper than any competitor does, you'll end up with a lot more big wins and far fewer losses. In this chapter, we'll study real-life examples of understanding customers' people, processes, and technologies, as well as how you might misunderstand them. My varied experiences will serve to help you put this essential aspect of magical selling into practice.

Let me demonstrate with a visual from the customer's perspective.

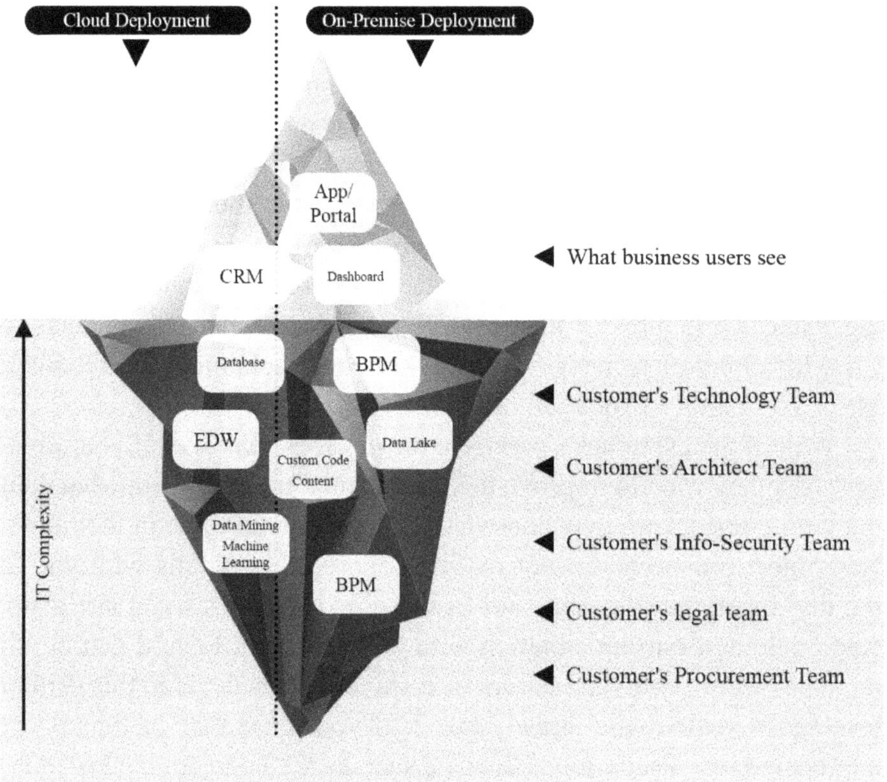

Figure 2 The Sales Iceberg.

With the enormous number of startups all over the world, as well as new products launched by large vendors, you have to jump through a lot of hoops to engage the customer early on in the buying cycle. In today's virtual selling environment, now accelerated by COVID-19, you need to plan to educate customers at least 60 to 70 percent on your product or service before they get in touch with you for the first time. Too many startups spend precious time in long negotiations. There is only one goal if you are a startup: get as many new logos—that is, as many new companies as your customers—as possible. There is no better pitch to get to the table than rattling off a list of customers that bought from you in the recent past.

Think of selling BMWs, Mercedes, and Audis. Now think of selling Avis, Hertz, and other rental cars. Then comes ride-hailing services. The base product is the same, an automobile, but it is the consumption pattern and therefore the business model surrounding it that evolves. The agility of sales cycles depends on your solution and the customer's readiness to adopt it. Simplify the process of sharing information within your team by making a living account plan to expand your technology footprint. Is it hard to get all this information? Yes, but you have to start with what you know.

Start to know your customer with what I call DON Strategy.

D—Defense Account/Territory/Market

O—Offense Account/Territory/Market

N—Net New Account/Territory/Market

A **defense account** is an existing customer, usually (but not necessarily) a high-paying one. Your wallet share is usually bigger than your competition's.

An **offense account** is one your competition has a strong hold on. Your wallet share is much smaller or nonexistent compared to your competition. This is usually a hostile account. You have less intel and fewer champions supporting you within the customer's company.

A **net new account** is nobody's stronghold. It's simply a new account without any history. In some cases, your solution may be new to the market.

There will always be borderline accounts that seem to belong to more than one of these categories. Don't fret over minute accuracies. The process outlined in this book is applicable to all three account types. More of this in the 4th Act, "Win with a Plan"

UNDERSTANDING THE CUSTOMER'S PEOPLE

A common ask of salespeople is to find out what your customer's Horizon 1, 2, and 3 outcomes are. This is another way of saying short-term, medium-, and long-term outcomes.

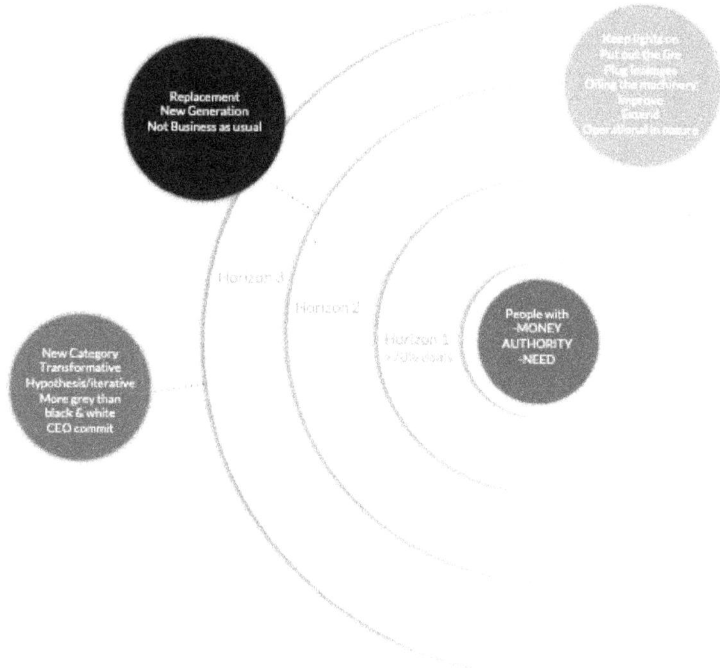

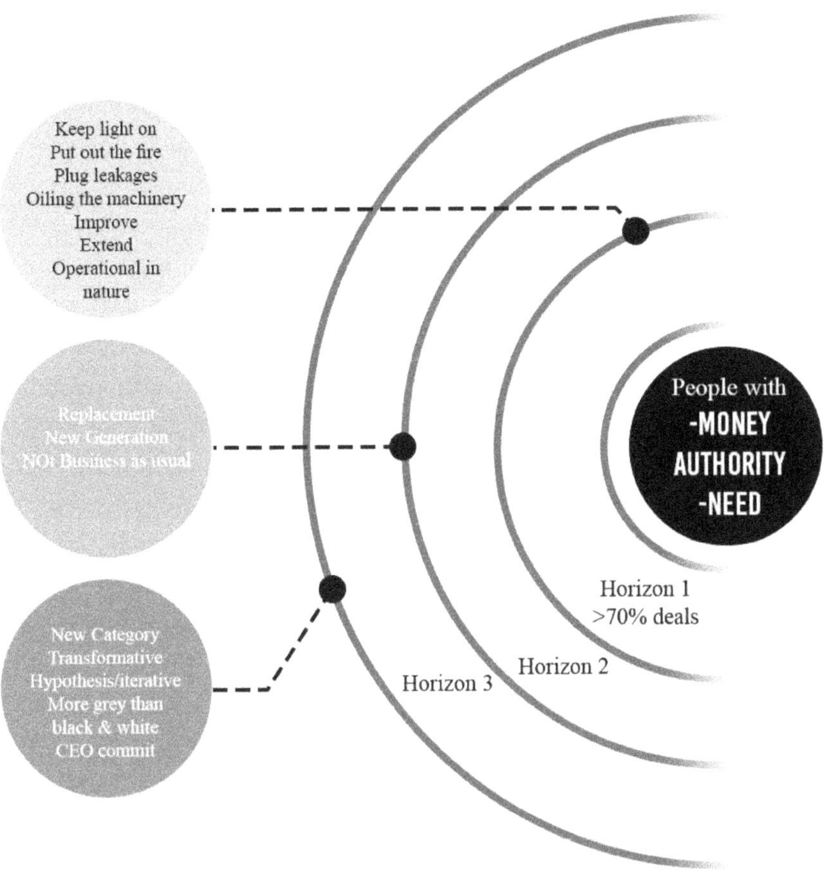

Figure 3 Horizon 1, 2, 3 Outcomes.

A very effective way to understand a customer is to think like the CEO of your own company. Start by listening to your CEO's every message and practice till you can repeat it verbatim. Whether you know this or not, great salespeople share several qualities with great CEOs, such as simple, powerful messaging; high energy levels; being comfortable operating in hazy situations; and the ability to energize people. Think like a CEO. Even if you've never been a C-level executive yourself, you're already the CEO of your own life. Think about it. We all have finite time and energy to realize our potential. After all, that's the essence of enterprise sales: leveraging everyone's time and energy for the best possible outcome.

Pay attention to the kind of people. Every department has Horizon 1, 2, and 3 thinkers. If you know your own value proposition fully, you'll find it easy to pace and lead the conversation with each of these horizons. Get help from specialists in your company.

Spend time learning everything you can about your business. Luckily you can do this any time you are free—weekends, the middle of the night, whenever. Pick a few mentors within your company or outside of it to learn more about areas you're new to such as products, the industry, technology, and the territory. Learn how to connect the dots between your solution and the customer's desired outcome in a measurable way. If your sales content doesn't simplify this, don't waste time criticizing the content; just go to someone who can clarify and simplify those for you.

I once worked with a sharp solutions person who was the vice president of solutions consulting, APAC. Often he would tell me about our new products and how customers could use them so my team could start selling them. On one occasion, he was describing some features that sounded like they were written in another language. I asked him to repeat himself. He elaborated, but I still didn't get it. I asked him to explain again. Round and round we went, but I still couldn't make sense of what he was saying, much less share that information with customers to drive sales.

The VP got impatient. "What's wrong with you?" he asked me. "I'm just not sure why this is so difficult for you to understand."

"I think the better question is, what's wrong with *you*?" I replied with a laugh. "We need to be able to simplify this in a manner that the average person can understand."

We went back a long way, so our back-and-forth wasn't personal. As I continued asking questions, he finally found a way to simplify the product specifications so I could understand.

Customers want to talk to knowledgeable, credible subject matter experts. After I finished a twenty-minute discussion on how big data can help improve supply chain efficiency with the CEO of a consumer goods company and his head of strategy, the head of strategy looked at my business card and said, "So we have to reach someone who is a senior director if we want to have a meaningful conversation with your company representatives." He was referring to other meetings he had had with my colleagues in which all of them were focused on how good our drillbit was instead of discussing the value of solving a supply chain problem for the customer.

UNDERSTANDING THE CUSTOMER'S PROCESS

Knowledge of the customer's evaluation process, decision making criteria is a fundamental part of selling. Who handles the paperwork, the approval process, the committees, the management, signatures and so on? Get this information as fast as you can.

Companies prefer vendors who can solve *all* their problems—the classic "one-stop shop"—with the clear benefit of a lower cost and reduced complexity, all while ensuring they don't feel "locked in." Follow the process I outline, and you'll become that vendor for your customers.

I say this with confidence because my efforts to understand prospects' processes (combined with people and technology) have multiplied company revenue ten to twenty times within a few years. I grew a territory with <$3 million revenue into a territory that generated $20 million a year. I transformed a shrinking territory that made $7.6 million a year into a $26-million-a-year territory in a mature, slow-growth market. Both were clear cases of taking market share from the competition by "sweeping the table."

Let me share another story. Customer was our stronghold account in Telecom vertical. We were selling big data technology before most companies were buying it. At the time, we weren't known for big data, and the market itself was nascent. This deal was a seven-figure opportunity, which was unusual for us in this product set. There were only a handful of such deals globally, and this surprised even our executive management. So, how did we engage the customer, understand their process, and make a deal?

We knew from experience that customers had 2 parts related to their procurement. One was permission to approve the purchase requisition followed by the Purchase Order generation process. These two happen roughly 2 weeks apart. A deal was not considered in the bag until one vendor made it through both processes. In theory a deal can be swung in a different direction.

Our main competitor was a hardware company. Six months in, this vendor got the better of us, kickstarting the first of the two transaction processes. In my experience, 99.9 percent of the time, the vendor that gets the requisition gets the purchase order because most competition disengages at that point. The fact that we understood the customer drove me to keep fighting against 99.9-to-0.1 odds. It was clear that we missed the boat when it came to linking the real benefits to our offer.

Two of our sales directors suggested we should just leave this sale to the other vendor

We understood the inner workings of this account. I wanted to use that information to bring this deal back to the table. One baby step at a time. We're going to figure out a way to win. What will it take? What things do we know about this account that will allow us to position our deal differently? We just need to come back with a better proposition.

Dennis (our regional sales director) and Joseph, our country sales director, were both really good at what they did. If anyone knew how to find an angle to get this deal back on the table and give us one more shot, it was them. Even though the odds were very low, what do we have to lose giving another try?

Dennis and Joseph got back to their champion and worked their way up through a couple of meetings to earn a final chance to present the following day. There were no guarantees, but their CIO agreed to review our revised one-pager value proposition. We had until the following morning at eight o'clock. 18 hours to prepare.

Joseph, Dennis, and I finished out the night by converting the insights from our discussion into a compelling proposal. By seven in the morning, our revised presentation with the value proposition was submitted to the CIO, who called us a few hours later.

Joseph met with the CIO in person that afternoon for an in-depth discussion of our offer and its advantages over the other vendor. One conversation and one handshake later, the deal was ours. While our competitor was probably celebrating their presumed seven-figure win, we were moving from procurement to purchase, all because we understood the customer's process.

My point is, if you want more sales at higher deal values, dedicate yourself to learning the inner workings of your customer's business operation. You'll win more often than not, even when it looks like you've lost. The trick lies in proper deal inspection, either with your coworkers or with a partner you wish to work with. It is not easy to pull off these wins all the time. In an established business you may find it a bit easier to close even with premium pricing. I was used to that. But in introducing new technology the priority is that of more logos—more customer wins—not more margins. I learnt this valuable lesson the hard way, after burning my fingers on some important deals. The margins improve over time.

UNDERSTANDING THE CUSTOMER'S TECHNOLOGY

Years ago, I was assigned to be sales director in an underperforming sales territory. Revenue had fallen from around $11 million to $7.6 million in the three years before I took the position. It was now my job to find the problem and fix it.

I met with existing customers to get a feel for how their needs were or were not being met to plan my approach.

One of them was one of the largest telecommunications companies in the region, a big spender, to say the very least. They had just cancelled one of their support contracts with us. To lose a deal of that size meant our engagement wasn't good. It's easy to figure that out. They had a stated objective of exiting our technology; a big and growing company buying from every vendor and not from us meant we didn't understand the customer.

The customer was spending big on digital transformation so I started by making that account into its own separate territory. The potential was so big that we needed a dedicated person focused only on that account. After all, understanding a large company takes time and effort.

We hired a new rep for that position, and I briefed her on the situation. We wanted to know what's going on in different departments (the M in MAGIC). What processes they followed and what existing technology they used. People, process, technology. I asked her and the local sales director, her reporting manager, to build that iceberg for me. This was a long-term task. She started to plan the activities to generate results on the three-year horizon. The first year was unlikely to be successful in terms of commissions, but I was convinced that if we focused on the right activities, she would be making a lot of money for both the company and herself.

For the first eight months there were many, many discussions with the customer. Not a single opportunity came out of them, but we did get a great deal of understanding of how they operated. We realized that the customer liked to project manage their lead vendors in any project. They gave each lead application vendor full service-level agreements (SLA) to manage the underlying 3rd party technology (e.g., platform, hardware, integration, security). This meant 'one throat to choke' to ensure better Service Level but also an enormous loss in efficiency. Why? Because each application had to be sized upfront for peak load that happens only once in a while. Meaning they used about 30% capacity on most days. This inefficiency, coupled with the complexity of managing day-to-day operations, led to them starting cloud adoption. But the reality is, not everything can be sent to the cloud. What about the efficiency for the platform? That was our entry point!

Nine months after hiring the new account manager, we found our first opportunity. An incoming request from the company told us they wanted to take advantage of the solution they used to buy from us. The deal value was roughly $1 million if we applied their discounts. But now we were armed with new information—I realized the way they were purchasing was simply inefficient. Even though a million dollars is a lot, and we were hungry, it wasn't enough money. It wouldn't be smart to just give our old deal away.

We told the company's CIO that we would no longer be able to offer the previous discount. We highlighted many inefficiencies that could be fixed if they let us do the end to end architecture. We said we could easily bring in a 20 percent savings compared to what they were spending, provided they allow us to build the framework. We later called that initiative "Mission 20."

Our proposal was bold, but we were confident because we had taken the time to understand the customer. They typically bought a database, a network, or a storage application separately and gave full management responsibility to the application vendor. Since that was their usual process, why would they change it for us? We could standardize the platform across applications. The fact that we knew their level of inefficiency was the only strength we had.

Any salesperson with this level of alignment would have won this customer over their competition. We just followed the MAGIC system when others did not. Anyone who diligently gathers data, turns that into information, and gains knowledge of the account can translate it into a material impact. After all, an impact of 20 percent cost savings across the board is every CIO's dream.

"Raju, if it goes below twenty, I will hold you personally accountable," the CIO said.

"Fair point," I said. "And in return, I'm asking you to be accountable for allowing us to transform your entire platform footprint for all applications."

The CIO called his procurement team to hold off on purchases from our competitors. There was no doubt that our products were top of the line. It was just that, when they bought a piece at a time, our products *seemed* very expensive. But collectively, the products could be assembled in a much more cost-efficient way. You see, "expensive" is a relative term. Getting perspective lets you sell stuff with a list price much higher than your competitor's but low on total cost of ownership. To outsiders, it looks like magic.

Within a week's time, the CIO called my colleague Joseph because he couldn't understand why, of all companies, we could get him 20 percent savings.

"Are you really sure you guys can save us twenty percent?"

"Yes," Joseph said. "We're sure. We've done it before with other companies."

"OK," said the CIO. "If you can do it, I want to add ten more projects to the one we've already discussed. Can you come in and take a look at those?"

That CIO ended up asking us to take over a total of thirty-three applications. We learned that they had a $24 million budget over a five-year period for things like replacing servers and technical support. We ran the numbers to do everything 20 percent under that budget.

Instead of closing just one deal, eighteen months from the day of that last call with the CIO, we ended up closing $14 million in net new business across multiple deals to support their data center transformation. The entire organization standardized on our platform, which was thought to be impossible at the time. Our account manager increased her pay by 450 percent that year and for two years afterward. Our efforts also stole market share from competitors, cementing us as the biggest player in that space.

This success story shows the importance of understanding the customer's technology better than anyone else. By aligning with the final outcome of Mission 20, we took unit pricing off the table. The technological benefits were connected to results. Our competition couldn't.

This was such a successful turnaround that the account manager gleefully told me later that the CIO shared with her what he told all other incumbent vendors: "Why do you guys come to me asking for projects? Why can't you be like our company? Come in and create a project that has material benefits and take the money?"

One final note: This account was an "offense account" with the customer planning to exit our platform as a part of their data center transformation when they ended up replacing our competition and expanded our footprint by five times. And achieved 20% savings!

WHEN THE CUSTOMER FEELS MISUNDERSTOOD

This commonly occurs when the sales team aligns only with the team they're comfortable dealing with, such as the technology team. Not everyone knows how to speak to the business team that ends up using the solution or vice versa. Let me be clear: people, process, and technology for *each and every department* need to be understood. Else you risk the deal. Just as I can tell you many stories about understanding customers from these three angles, I can also report on the times I failed to do so—and what we did to turn them around.

Our platform sales team rarely engaged with the customer's business team to close analytics and data warehouse deals. Our engagement was only with the IT team. At the time, we approached the customer lifecycle management (CLM) team of a large subsidiary of a European brand for an analytics and data Warehouse deal. But, like I mentioned, we didn't sell to their business teams.

One of our best solutions architects, Joshua, and I went in to present to them, prepared to talk about how we engineered the hardware and software to work together in a very efficient way. We lost the audience within the first few minutes because we were too technical. They were interested in how they could meet their KPIs. The customer was polite, but it was time to leave.

After that meeting, I realized we didn't have the subject matter expertise to have meaningful conversations with business users. We knew our product (the drill bit) but had no clue how it connected to business for the CLM team (the hole). The unsaid question from the customer was: “Yes, you guys have super-duper technology, so what?”. We needed somebody who could translate our infrastructure’s features into its capabilities, and those capabilities into KPIs. The country sales head suggested that he knew a guy in our French office. We reached out to that product marketing director for our communications global business unit. Remember, your colleagues are often the secret to getting a customer’s attention. An ever-important aspect in winning is also to catch your competition off guard when it comes to your next move. My newfound knowledge of our KPIs was a way to position ourselves on higher ground than our competition. The competition wouldn’t expect us to lead with KPIs and subject matter. If we did that and did it well, it would be easy to take the pole position ourselves.

Armed with the knowledge, we reconnected with the company. Our competition went about business as usual, emphasizing why they were the best choice. We broke from that approach during our second conversation with the CLM team. This time, we led entirely with how we could help them meet their KPIs better than any other company. The talk of metrics kept everyone’s attention. The presentation and follow-up questions ended up filling an entire day. By four o’clock, we hadn’t talked about the platform features once. During this evaluation that lasted several weeks, we successfully earned a technically superior score (T1), which was our 1st goal. These T1 rankers mostly end up winning the deal while the rest wait for the T1 to falter.

When the discussion turned to pricing, they told us our quote was much higher than the competition’s. For reference, that deal was worth $5 million to $6 million of mainly software and some hardware. Typically, customers bought each product from a different vendor. We provided all of these, and our hardware was known to be more expensive than what our competition offered. Because our competitor didn’t make any hardware themselves, they outsourced the cheapest solution available to maximize revenue. We used that to our advantage.

“We’re putting together many different things with this deal,” I said. “My team is only being paid on a portion of them. Still, we put together the entire deal based on what will work well for you. Unlike our competitors, I'm not going to choose the cheapest solution for you because that's not in your best interest. Sure, some components can be sourced a bit cheaper, but only one of us is thinking about your KPIs and project success. We did not compromise on quality. It's your decision what is more important to you."

We won the deal. That one sale led to other subsidiaries buying similar deals. We led each of them with success metrics.

When you dedicate yourself to understanding customers, and align your resource allocation, you look like a magician. You rescue ruined deals at the last minute. You do the impossible with ease. It's not magic. It's engineering. You get to engineer more wins, more commissions, quicker promotions, and even a new job if you want. Just like in the case of well-engineered products, a well-engineered sales process is not about how good it looks; it's about how well it works

RETHINKING "NO CHANCE" DEALS

By this point in the chapter, you may be second-guessing your past sales efforts. How many deals that you thought you couldn't win were actually wide-open opportunities? Did you just need deeper customer understanding? You may have assumed your chances were so low they weren't even worth reengaging after the first no. You know better now, but many of your competitors still do not. For you, a "no chance" deal is simply an opportunity to learn about the customer's people, process, and technology—and use that knowledge to make a deal. Let me give you another example.

This is a combination of an **offense** and a **net-new deal:** offense because we were in a country where the competition had a much better revenue share and a better ecosystem of skills, and net-new because the product this customer was requesting was new.

A retail company deployed a merchandising solution. Five years later, they had grown so fast that they had outgrown the enterprise resource planning (ERP) functions of that merchandising solution and needed a more robust software to manage all business operations from one dashboard. At the time, my team and I were not responsible for selling ERP, but every ERP needs a database. That was where we got involved.

A market-leading ERP vendor quoted $399,000. The retail company's CIO told us that they would consider our ERP modules solution if we offered more value at $250,000.

If we took the Database/ERP route to win, our chances were low. The only way to increase our chances is by working on the final business metric that their CEO cared. We shifted our pitch from ERP modules to our industry expertise: the retail global business unit. Why? Because customers are really solving a business problem, which was to lower inventory costs.

We went to Aylwin, the retail business unit leader in charge of Southeast Asia, and worked together on how to win this account for us. We had to get the customer to tell us what they really wanted—in this case, inventory cost reduction. Then we showed them how to save 2 to 5% of their inventory cost with our solution. In parallel, our ERP competitor had poisoned the well, so to speak. We had many products and therefore many teams and were known in the market to have over a dozen people from our sales team that overwhelmed our customers. We knew that, and we expected that they would use this against us. They had warned the retail company owners that we were too complex and difficult to work with. We used this seeming weakness to outsmart the competition. We had our retail account manager lead all future meetings with the retailer's leadership team. He would be our single point of contact. Just him. Having one person handle the account would make us out to be a one-stop shop and hit the credibility of our competitor. Anything and everything the account manager needed to know about databases, business intelligence, middleware integration, and database security, we filled him in on. If we executed well, there was no chance that our competitor would win.

After eight months of meetings, presentations, and demonstrations, we closed an end-to-end retail merchandising solution with a full platform deal worth $3.1 million—much higher than what we initially expected given the CIO's stated budget. It turned out that our competitor, the industry leader, had offered a similar package deal for $2.2 million. Usually quote differences that far apart motivate a prospect to negotiate hard with the higher-priced vendor, but we didn't need to.

In order to run that sales cycle, we needed someone with that industry knowledge. Aylwin was that guy for me. He was from a different team and didn't report to me at the time. Partnering with someone with skills not available on your own team is essential. That partner may be from a different department or from an external partner organization.

One more example, from when I was a sales rep. The customer was a well-known media organization. Early on in that eight-month period when I was selling databases for the company's advertising system, the CFO told us that they were meeting with our competitor separately for their integration/portal initiative. He would not allow us to participate in those discussions because we were not well known for that. Basically, "Thank you very much for inquiring, but we don't have time to restart the search."

A week after we closed the database deal through a systems integration (SI) partner, the CFO called and blasted me for not giving him the best price. He said some other partner approached him with a better discount than he had paid. I couldn't understand why he was furious with *me*. I wasn't the one who had negotiated those partner discount deals. I told him it simply wasn't possible for me to rebook his deal. I knew they were finalizing the portal soon, so I suggested that if he allowed us to bid on the portal deal, he could either choose us at a lower price or negotiate with the winner, using us as a competitive threat.

"It will be a win-win for you," I said. "Either you'll like our product better, or you'll get a better price from the other vendor due to competitive bidding."

The CFO liked the idea. We had to move fast with our proposal because the other vendor was already engaged and trying to close the transaction. The good news was that they had been working on the deal for more than a year. The media company basically handed us their entire requisition, making it almost too easy to put together a competitive proposal of our own. When you make it clear that the customer wins no matter what, they find it very hard to walk away.

Two days before the deadline, our proof of concept was ready. The CFO was shocked and impressed that we could finish the POC that quickly. We won the transaction for a totally new product. Essentially, a well-known media company considered us better than the market leader, even though they were literally the first one to deploy that newly launched product in all of Asia Pacific.

The deal stunned our competitors and pleasantly surprised our senior management. That such a deal could have been closed entirely on the phone was impressive. This was a $100,000 deal. What we had done probably looked like magic, pivoting our database offer and sweeping the table. Remember, one person's magic is another's engineering. I had established my credibility as someone willing to set up a win-win scenario for the customer. Customers appreciate that, but you can't do it unless you understand the customer, their approach, and the problems associated with the product they're buying.

You'll be surprised how often understanding the customer better than competitors exposes those competitors as unprepared. They won't know what happened. Meanwhile, you'll be closing big deals and earning big commissions.

If you're David facing Goliath, use these tips:

- Find out if your competitor is asleep at the wheel. You'll be surprised how many vendors are!
- Make a list of things your competitor's salespeople would probably say about you, then do exactly the opposite. Tear holes into their credibility.
- Avoid statements. Demonstrate flexibility and adapt at every step. Remember the magic words "convince, or be convinced" for a frictionless deal progression.

- Don't talk too much about competition. Be smart and sensible, sharing with customers what they don't know instead of what they already know.
- Remember that the customer's motivation is to reach their KPIs better and easier. They're not wedded to any specific vendor if they find a low-risk alternative.

A combination of these will close some distance. It may not be enough to win the deal yet, but you're sure to lead the sales cycle. Look back at the improbable deals you either won or lost. In both cases, the winning team most likely executed the above points better than the loser.

FIVE QUESTIONS TO UNDERSTAND ANY CUSTOMER

The examples above have given you a look into real-time customer understanding. Now I want to take you behind the scenes. What questions did I ask the decision makers in these situations? You're about to find out.

There are five simple questions that enable you to understand all three aspects of your customer, their people, process, and technology. These are exploratory questions that allow you to discover the customer's real requirement. Questions like, is there a budget? Who is the decision maker? These are table stakes. They are your first step to understanding your customer.

From this day forward, you're going to ask all prospects these five questions. You may not get all the answers you need overnight, but over time, you'll equip yourself with more knowledge of your customer's situation than you thought possible. How much time, exactly? Some deals can take months, others only weeks. In rare cases, you can close in days. Regardless of how long the understanding process takes, it's worth it. Seek out the answers, build your offer around them, and watch as your closing rate and deal value skyrocket.

QUESTION #1 HOW DO YOU MEASURE SUCCESS?

How does the customer determine if the project is a success? This could be in terms of time spent or saved, effort invested, complications simplified, or money saved or earned. This will help you to align (A in MAGIC) with them. Each team (or its members) will have their own measure for success. What is the quantitative measure they are aiming for with this project?

QUESTION #2 WHAT IS IT NOW?

What is the measure of that success today in time, effort, manpower, or dollars? Or is it a brand new initiative never tried before?

QUESTION #3 WHERE DO YOU WANT TO END UP?

Whatever the end goal, find out what they've tried to get there. Be open to the fact that customers may or may not have thought through their success. Sometimes customers don't want to expose themselves to a certain improvement. They may be trying a new solution to simply improve a troubled situation rather than solve an expensive problem.

Sometimes it's simply about shifting work from doing it yourself to getting it done elsewhere. Talent availability and retention can be big problems in many cases. There are many instances where companies outsource positions or services so that they don't have to hire, manage, and implement solutions themselves. "When you can buy milk from a nearby shop, why would you want to buy a buffalo?" If you have buffalo at home, you can argue that the milk is free, but you have to clean the buffalo, feed it, take care of it, and so on. Your job as a salesperson is to understand the "buffalo" of your customer. Use this metaphor carefully.

Otherwise you may imply that your customer's job is redundant. If you do that, don't expect to win. Pack your bags and leave. You're not saying anyone's job is redundant—you're pointing out that their buffalos are redundant.

QUESTION #4 IF YOU MAKE THESE CHANGES, HOW MUCH WOULD YOU SAVE?

This question is a variation of the previous one, but it's critical to ask. You need to know the quantifiable benefit of your proposed solution. Again, prospects often won't answer this right away. Customer may not even know.

QUESTION #5 WHAT IS THE BENEFIT OVER TIME?

Question #5 builds on the last question, extending the quantifiable outcome over the next three to five years. If your solution solves a customer's problem to the tune of saving $X per year, that's $5X over the next five years. A projection like this will give you a sense of how much the customer is likely to spend.

If we step back and look at these five questions carefully, we notice they are based on first principles. Anyone selling technology should know that technology can create a bottleneck or destroy a bottleneck. The heart of the opportunity is nothing but the application of this in the context of your customer's need.

What is your role? How does your customer plan to deploy your solution (or that project)? Is it to destroy an existing bottleneck or create a new one?

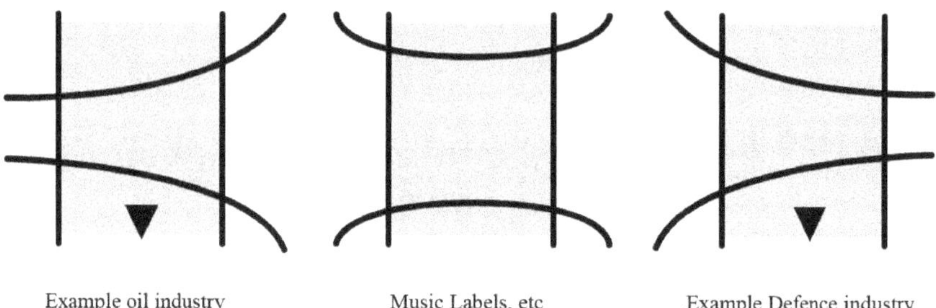

Figure 4 Solution Deployment Examples.

Your ability to articulate how your solution offers measurable benefits puts you in position to win the deal. This picture helps you understand the context—the C in MAGIC—clearly.

It's Got to Be a Two-Way Street

Getting to the root of the problem or the heart of the opportunity is a two-way street. Don't compromise if it is not. Your organization needs to have presentations ready with the same quantifiable messaging on how your product helped other customers. If your company can align its internal messaging with your customer's answers to the above five questions, you can save a ton of money in efficiency. And guess what? Many customers out there may not have a targeted measure for success. If they don't, you might end up looking like a solution-selling or value-selling person. That's the magical outcome even if you weren't known to be capable of those skills before.

Allow me to give an example of how knowing this exercise is a two-way street allows you to compete better. A bank in Asia was evaluating our engineered systems. Back then a competing product was better known and more widely used in that market. One of the differences was that the competitor's chips performed better than our Intel-based ones. The CIO was getting into those nitty-gritty comparisons based on transaction processing performance (TPC) benchmark reports. It was really one of the many objections against doing business with us, and was classified as a genuine issue in my objection handling techniques (i.e., objection types I will talk about in chapter 6).

But knowing the above information allowed us to put the question back to the CIO: "Sir, we can go into each and every part of the system and find a third-party report to support one vendor or the other, but the reality is that when you have different products from different vendors, you have to assemble them together in your data center. How will you troubleshoot when something is down? With each vendor pointing fingers at some other product being the root cause. And if that doesn't get solved and you don't tally the books of the previous day's transactions before the bank opens, who will be responsible for the loss incurred by your bank? With our engineered systems, it's easy to troubleshoot and not waste valuable time and effort. We get to fix it. Your measure of success is really ensuring the bank opens on time every day. How does it matter if one chip is a few percentage points better than the other in a standard third-party test? Is that really the bottleneck?"

The CIO thanked us for reminding him of his main purpose and never brought that up during the rest of the sales cycle. You will be surprised how customers will drop even their preferred vendors quickly if you find a way to align well with their success metrics. The skill is to get customers to share that information. The rest appears as magic to the untrained eye and to whomever it belongs, be that your boss, your partner, your competition, or your colleagues.

In another opportunity a CIO of a small bank wanted only the production system and had no need for disaster recovery. He had a low budget. We asked him why he thought a disaster recovery system was not required and how he planned to minimize his company's downtime if such a situation arose. I suggested to the CIO that he present to the board what the risks were for not having a disaster recovery plan and then ask for their explicit approval so they wouldn't point fingers at him later. When he presented that, the board actually approved the disaster recovery setup for our solution, and we increased our deal's value by 75 percent.

Do you see how connecting to the business outcome in both cases triggered a deal expansion and increased our chances to win over our well-known competition?

A SHORTCUT TO UNDERSTANDING CUSTOMERS: ARE YOU SELLING ASPIRIN OR VITAMINS?

It's easier to sell aspirin than vitamins. Aspirin is a commodity. It flies off drugstore, supermarket, and gas station shelves daily. Nearly everyone has aspirin in their home medicine cabinet. Premium vitamins, however, are only sold to a certain clientele. These consumers have to visit specialty markets and health food stores to find the entire selection.

Whenever you're selling, keep this shortcut in mind: If you sell aspirin, your product is a regularly budgeted activity. Aspirin has many obvious uses (e.g., headaches, fevers, injuries). Think Horizon 1 (in general). Aspirin transactions are simple, frequent, and completed quickly. You can't increase profit margins on commodity products, but you can give someone a winning deal on aspirin and put in the extra effort to show your customer how it benefits them.

Every once in a while, an aspirin customer decides they want vitamins instead of or in addition to their usual pain relief. Because you worked hard to earn their aspirin business, they're likely to buy your vitamins—even if they're more expensive. At that point, pricing is of little consequence. It's about the relationship. You understand the customer, and they know it.

Also, when you put in the extra effort and give someone a winning deal on aspirin, they'll remember you. Then, when the day comes that they need vitamins, they'll come back to you first. It's not really about what features and functionalities your vitamins offer. Most of these products are basically the same. Like I said before, I could buy a Honda or a Toyota. They're both decent cars. The key is putting your products in context for the customer. Show them how they specifically will benefit from your product.

ADAPTING TO THE POST-PANDEMIC REALITY OF VIRTUAL SELLING

The novel coronavirus, or COVID-19, pandemic has disrupted both buying and selling behavior big time. I prefer to address this from the customer's viewpoint and build from there. Here is how I depict it . . .

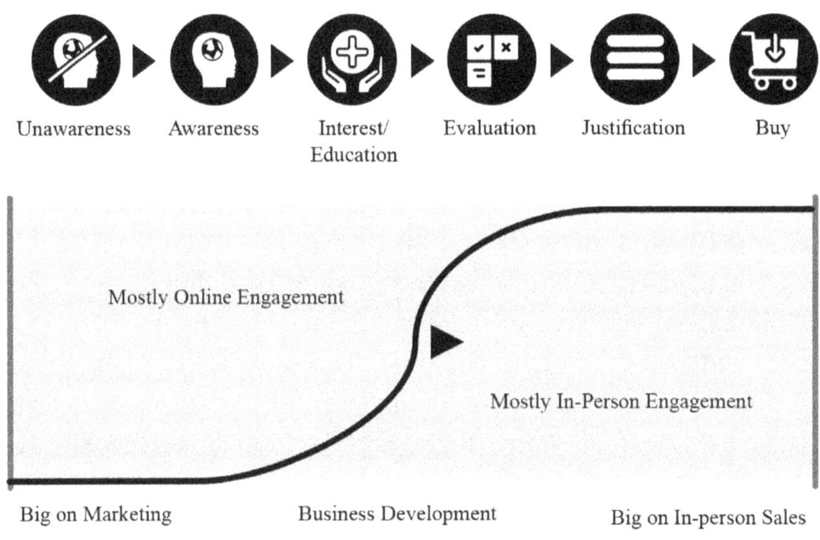

The curved line depicts point in when you usually start in-person engagement with clients. We meet some clients early and some later in the buying cycle. You will see that curve moves to right all the way until almost all the sales happen remotely.

Figure 5 Client Engagement Behavioral Shift.

This pandemic has been the biggest catalyst for digital transformation in all brick-and-mortar businesses. When the dust settles we will not return to our previous way of working. If you are waiting for the good old days to return so that you can resume sales meetings, good luck to you. That ship has sailed. It was fun while it lasted.

Customers are now changing their buying behavior and are more open to virtual evaluation even for large projects. Even if you are in field sales and know many customers, if customers get used to buying virtually for over a year and you get used to selling virtually over a year, the process changed forever. You will have to make smart changes in order to adapt.

I have experienced this before. After spending the four and a half years of my career in field sales selling applications, starting up new territories, and new technologies like hosted solutions (the initial name for cloud-based services) and e-learning systems, I got a job at Oracle selling products virtually on OracleDirect. I had no idea what remote sales were, but I didn't hesitate because Oracle had doubled my salary! What was there to think about? I knew I'd adapt however we needed to sell. We used the internet and telephones and webinars to replicate the effectiveness of selling in person. Remember that we were only selling remotely while our competition could meet with customers. The division was meant to take care of transactional selling, but as things turned out I would always focus on pushing the boundaries of value selling and solution selling. I had to rely on partnerships, devise a way to monitor them effectively, and get them to work for me. They were my eyes and ears the bulk of the time. That was where I strengthened my ability to delegate any transactional part of the sale to my partners and leave the not-so-easy ones for myself. It was a double benefit: Keep the deals flowing on one hand while freeing my time to focus on deals that others wouldn't think of doing. The best part was that I needed to be better than my competitors' field sales team if I wanted to succeed. I spent four years and got to manage a team and train them on my way of doing things. This was where I sharpened my ability to listen more carefully, find ways to get intel, get customers to talk more freely to me, and negotiate all while over the phone. The differentiator was clearly outcome-based selling tailored to the context of every customer. I never behaved like a telesales rep because of my field experience. I was clearly practicing the magical sales methodology you are holding in your hand.

Context, planned incremental progress, quick turnaround times, presenting outcomes unlike most competitors, refusing to engage only on features and functionality, internal teamwork, and sharing win plans helped me make an impact with customers and partners. That experience early in my career helped me adapt to any situation. When I got back in the field four years later and sold to faraway places, I pushed hard to have three times the meeting productivity than someone would if they lived in the same city as the customer and could meet with them more frequently.

There are not many people out there who could get solid results on such a spectrum of roles. At its core, magical selling by design is built to adapt and therefore built to last. What better situation to try out in the biggest disruption in a century? With today's technology we can nearly replicate the in-person sales experience. How we use it is up to our individual skill sets.

Tips:

1. Understand the product fully with the ability to measure the customer's outcomes.
2. Prepare your own content and your company's content to align with this.
3. If the context is a mature customer in the framework of the customer's purchasing journey, simply align with their outcomes.
4. If the context is a not-so-mature customer, lead and set the metrics and help them win.
5. Capture everything in one place, your CRM, and inspect it daily with your team.

Traps:

1. If your training starts and ends with features and functionality and your tests or certifications end with that knowledge, be ready to be surprised by deal losses.
2. Get mentors to help you understand the customer's metrics.
3. Make a "So What?" strategy. If there is a new feature, find out what it is and ask, "So what?" as many times as needed until you get to the final quantifiable benefit—time, effort, SLA, price, results, whatever that might be for different individuals.
4. Avoid taking the position saying "this is our standard." No one has time for inflexible companies, so the customer will jump

off at the first opportunity. Weekly forecasting is a phase. Real time it is.

5. If your internal system doesn't have smooth collaboration facilities, find another job.

YEAH . . . BUT

It's a legitimate thing to say, "Yeah . . . but":

- "I've never done it before. I don't know how to."
- "What does good look like?"
- "Many of my customers don't want to share. I've tried!"
- "I'm doing well already. Why would I change that?"
- "No one in my company does this. My boss doesn't encourage this."
- "My customers don't do this themselves. How will they share with me?"
- "My customers don't want to tell me those answers."
- "What if I'm unable to carry out the conversation? What are the possible objections that I need to be aware of?"
- "The heart of the opportunity doesn't apply to my product or service."

All these objections are a function of the customer's maturity level, your own maturity level, and the strength of your relationship in that opportunity. The two guiding principles that I find effective are:

For customers or colleagues:

He who knows not, and knows not that he knows not, is a fool; shun him.
He who knows not, and knows that he knows not, is a student; teach him.
He who knows, and knows not that he knows, is asleep; wake him.
He who knows, and knows that he knows, is wise; follow him.

A workflow-based approach (if you prefer) to navigate:

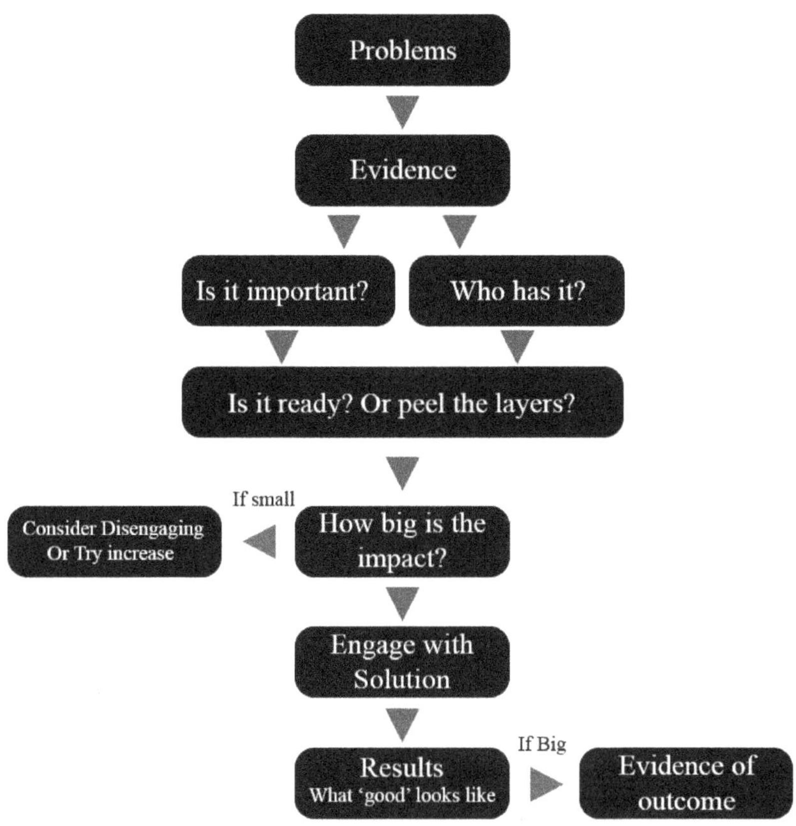

Figure 6 Objection-Handling Workflow.

THE MAGIC FORMULA FOR FRICTIONLESS NEGOTIATION

Just like selling, negotiation and objection handling is an inherent part of your everyday work. All of us need to improve our negotiation skills in both our personal and professional lives.

At my first sales job back in 1999, I attended a training seminar on negotiation skills. I read several articles and a couple of books on negotiation, but frankly, they looked like tricks to manipulate while I preferred authenticity in a more holistic way. One might have gotten good results on a few deals with these negotiation skills, but I found many of them to be short-term oriented with risks of compromising your credibility in the process. Not a good trade-off. I prefer creating strong relationships based on results we create for a win-win.

There were about twelve of us at that training, each from a different company. The trainer asked us if we thought we were getting paid what we deserved. That was a smart question to start with because everyone paid attention. Everyone, including me, agreed we didn't get paid enough. Everyone wanted to negotiate a higher pay. Then the trainer said something that stayed with me: "In the short run, you get paid what you negotiate. In the long run, you get paid what you deserve." When all moving parts are well aligned and well lubricated, the friction disappears. The four-word lubrication formula I experienced to sail through difficult circumstances is "convince or be convinced." You'll notice that customers will help you win deals, as a result.

DOCUMENTING YOUR CUSTOMER UNDERSTANDING

In enterprise sales, limiting the understanding of a customer to yourself is like sabotaging your own success. What's the point if everyone on your team doesn't understand the deal as well as you do? How do you ensure they contribute super effectively to help you win? Get into the "CEO mindset." We are in the age of digital selling, so there are many easy ways of capturing information. The pandemic only accelerated that trend. If your company is not making it easy for you to capture information, there are many tools out there to use, from account plans to opportunity plans to closing plans and even general activity on an account. Any company that requires the sales reps to replicate a documentation task in more than one system is really wasting a lot of money under the pretext of saving money. Companies are saving so much money on travel budgets. Why wouldn't they spend a part of that to automate and improve sales productivity?

1. **Your CRM system is your bread & butter:** One aspect of capturing details effectively is not needing to do it more than once. Some salespeople are sandbaggers, so they avoid adding specifics in the CRM. That's not smart.
   a. Your CRM gives you a chance to demonstrate the best practices in good times and opens up opportunities for career advancement.
   b. It will help you defend yourself better when things are not going your way because you will be able to show the effort you made in building a pipeline and deal progression.
   c. It protects your boss from needing to spend precious political capital to defend you in times of need. You would want them to use it to increase your pay or promote you instead of explaining that you work well if someone doesn't know you or work with you regularly.
   d. It gives an opportunity to approach seniors or leaders to coach or mentor you and to course correct your work for more effective use.
2. **Activities:**
   a. Forecast from inside your CRM. Forget slides.
   b. Prepare account plans or territory plans or close plans, again from within your CRMs. There are tools you could use these days. Virtual selling is bound to accelerate this.
   c. Incremental activity simply takes minutes. I'm not recommending which tool you use but I had a chance to observe Asana, Trello, fireflies and many more are all directed at

simplifying your administration time, freeing you up valuable time to keep talking to customers.

d. Learn to leverage to get internal negotiations going on who needs to work. Meet customers wherever they are in the buying process and their level of maturity. This is the Inspection (I) part of MAGIC. It might be less interesting but very effective.

Sales is like entrepreneurship. Everything required to persuade prospects to strike a deal with you is your responsibility. Even if you want to make sure your sales team knows all the customer's information as well. The buck stops with you. Fortunately, the strategies you learned in this chapter empower you to get the highest return from your sales activities. At every step in this process, your one and only focus is to understand the customer's people, process, and technology. Do whatever you can to learn about these, and you will have the advantage over bigger, cheaper, more established competitors nine and a half times out of ten.

Having the advantage will not ensure your win. I've seen many transactions fall apart in minutes despite the sales team perfectly understanding the customer. What was missing? That all-important second element of magical selling: Alignment. The fact is, it's not good enough to understand a customer if they don't *realize* you understand them. So how do you do that? Simply turn the page to find out.

ACT 2

CHAPTER 3

Validate That Customers Know You Understand Them Better than Anyone

"It's not sustainable if it's not a two-way street"

It is one thing for you to claim you understand the customer. Happens all the time. It's entirely different for the customer to say you understand their needs better than your competition. To get there is simply a continuous validation process to ensure that you get that position. I have seen any number of instances where deals are won or lost at this point. Seemingly smart and experienced reps have a problem making inroads because customers don't share enough information. If that's your situation, you should ask yourself or anyone on your team engaged with the customer who is not getting traction the following question: Does the customer think we understand them better than our competitors? See if answering that gives you clarity on why you won or lost. I recommend that you find ways to validate this at a deeper level to set yourself up to win. It eliminates the assumptions and guesswork to a large extent.

I always heard the advice that salespeople need to be trusted advisers to customers. That goal is clear. What is less clear, though, is how to become that trusted adviser. No one told me how to do that in a simple way. I saw people be trusted advisers to customers they had known for years, but I knew they built that trust over time. Many in-country sales teams have such a position. If you are a rock star on a certain subject, then sure, you can be a trusted adviser too. But my career path of being regional and in second-level management for many years meant that those opportunities were limited for me. So, over time, I developed this subtle nuance: What if I was meeting a hostile customer or a new customer for the first time? Was there a simple way to ensure that customers knew I understood them better than everyone else?

In reality, no customer ever told me in so many words that we understood them better than others. The message was always subtle. Sometimes it was revealed in their post-sale feedback. To a large extent this is just an observation of how customers respond to your Act 1. Three visible signs that a customer trusts you more than the competition are a) the customer lets you lead the conversation; b) they start sharing more information, effectively guiding you to a win, and c) they find it hard to negotiate with you. This is not a one-time occurrence but continuous in any sales cycle. A couple of examples. . .

A competitor pitched the data model for which they were known to a telecom company's general manager. =The model was like a preconfigured way of generating reports, dimensions, and KPIs in any industry. Many customers custom-build those, so the competitor had both "off the shelf" and "custom built" success stories. Meanwhile, ours was just a platform. We did our homework knowing from the ground team that the customer had developed over three hundred reports and were building more on the current platform, but the general manager didn't know that we understood. In our next meeting he said he was open to a data model. Instead of saying why a data model was not a good idea, which can become a theoretical discussion, we pitched it in a way that demonstrated our understanding of the customer better than the competitor.

We said, "Sir, we understand from your team that you have more than three hundred reports and KPIs already built over the last three years. Are you unhappy with those? Are the users complaining?" The

GM said no, that they were happy with those. We asked, "Then why do you need a data model? It does the same thing anyway." The customer looked at us as if thinking to himself why he hadn't thought of that before. The fact that we didn't have a data model didn't even come up. We simply demonstrated that we knew their ground reality compared and aligned with it whereas the competition was focused on hard-selling a solution, which may have been the best in the market but exposed their lack of understanding. We won that deal without giving any discount in the final negotiation. Our partner later told me that it was the first time he had witnessed a negotiation where we didn't give away a single dollar in the final negotiation and still won a deal in that account. To him it was magic!

This works internally too. You don't want yourself or your team to be on a wild goose chase. A few years ago I was having a coffee chat with our finance director. She was painful to deal with because the approval process was always tedious and time consuming, and most deals needed an exception approval those days. At the same time she was good at her job, with a reputation for immaculate paperwork, and I admired her for it. In that conversation she said something interesting. It was that she always prioritized my team's deals to be moved quickly over others.

That was a surprise to me. I asked her why. She said that the finance team knew that whenever my team put up an approval request they knew we would win, so she thought it was worth her time to work on our deals. The rest of the teams would eventually lose more than 50 percent of the deals for which they sought exception approval and so she felt like she was wasting her time. Imagine that. It is the responsibility of the sales team to avoid spending back-end resources if they didn't know if they would win or lose. My focus was always on establishing that our deal was winnable with reasonable certainty before involving finance or back-end deals. Presales and other GTM teams are fine because they help you in validation, but the back end should not be stressed unnecessarily. The trigger for me to start a process was when our customers validated that we understood them better than our competitors.

If you perform the first act of understanding the heart of the opportunity well, you have won half the battle. At least you won't be behind anyone and working toward proving that you can deliver those results. It's easy to say it, but in real life you will find a variety of customers. It is possible to pull off the heart of the opportunity if customers are mature and they know it. Even if they are not mature enough to set their expectations, it is possible to pull this off if you educate them. In both cases you have to engage with customers early in the sales process. What if you intercept a deal in the latter part of the cycle when requests for proposals (RFP) are released? What if the customer is hostile to you for some reason or already knows which vendor they are choosing before the RFP is released?

The reality is, customers usually put in a lot of effort prior to releasing an RFP. Vendors help by responding to a Request for Information (RFI) or educating customers by demonstrating the functionality of a product over a period of time. Some prospects also ask for a proof of concept to make sure they'll be satisfied with a vendor. Then they'll release the RFP to get the best price with equal or better functionality.

Remember, if you're not involved in influencing the RFP for long enough, your chances of winning are generally low. I'm not saying you should never respond to RFPs if you didn't influence them, but it takes a different approach to win such deals, like partnering with someone who has a better understanding of the RFP than you do. More of this in Act 4. For now, we'll focus on how to go about influencing the creation of RFPs, thereby increasing your odds of a win.

Customers appreciate being asked about their priorities. Simply taking an RFP and running with it shows an urgency that's unattractive. If you take the time to go beyond the formal document, preferably before the release of the RFP, you prove that you're an experienced salesperson who knows how to bring value to the table. There may be top-priority requirements the customer has that you don't know enough about. In that case, ask a subject matter expert on your team for more information. For example, a senior technology person could speak with the customer to establish an order of priority.

If you know your stuff well, customers don't mind giving you answers. Some may not be willing to proactively share information; others are mandated to *never* share a certain information; but if you ask, there's always somebody who can give you the answers you need. The more interested a customer is, the more information they'll be willing to give you, and the more counterquestions they'll ask. Answer their questions quickly to keep the deal momentum; strengthen the relationship, build the rapport, and become the trusted advisor.

That said, a few words of warning on blindly educating your customers: Watch out for customers who are the "see more" type. I know the CEO of a midsize company in APAC who turns up for any 'important' vendor and genuinely tries to understand but never buys from them. I had the experience myself, and I cautioned others on more than a couple of occasions when I heard they had met the CEO of that company. The company even released RFP and made it an in-depth engagement but didn't award any project to these vendors. This is why you need to qualify your leads before you put in a lot of effort. In the era of virtual selling it's possible for you to plan these as pipeline generation activities and allocate relevant resources to them. Read chapter 6 for more specifics on this. I'm not suggesting that educating customers is wrong, just that you should sharpen your lead qualification skills. A lot of customers say they want things done yesterday, but they don't do their part to execute rapidly. What they mean by "done yesterday" is that they want *you* to move fast. Really, they're asking you to *educate* them as quickly as possible about the right solution for their needs. And there are times they take that information to another vendor and buy it from them for half our price.

Don't give away too much information in anticipation of getting that yes. *It's OK to walk away*. Somehow trainers always teach sales people what to do, but not enough time is spent on when and how to walk away. The outcome of you following the advice in this chapter is you will know when to walk away. Keep that thought in the back of your mind at all times. Your job is to ensure you don't waste precious resources. If you sense your goodwill is about to be taken advantage of, the deal may not even be real. In that case, you're better off walking away.

Serious customers tend to give you additional intel and give you access to their decision makers and talk about outcomes. I realized that as long as we anchored effectively on the outcome they really cared about, they simply didn't care which specific technology could solve it. When customers know you understand the drillbit and the hole accurately, you will better control the sales cycle.

As you make an effort to understand your customer's needs and reflect that back to them, you'll encounter varying appetites for risk. A European bank wanted an outcome-based setup. It was the first time they were sharing sensitive data with external partners and as a result required regulatory compliance, but they needed to see if our solutions would work in their specific situation. It indicates the customer is serious about choosing you. They already believe you understand them better than other vendors; they simply want to validate that you're the right partner for them.

There was no direct competition for the deal. Our biggest competition was the customer doing nothing! When we finally got our first meeting, we were looking to get the answer for a straightforward question: "What is the minimum viable proposition?"

This question cut past everything we could have discussed, such as product features, advantages, and benefits. In other words, was it a real opportunity?

The response gave us a chance to align our pricing with their success metrics. We split that into an upfront cost so that we were not out of pocket and split our final success along a road map with ten milestones. This made us a partner. I see this kind of pricing discussion on the rise, especially with startups. This gives you a chance to plan for customer success upfront, which is pivotal to have stable growth in the crowded startup world. Watch your pipeline for clues on how to do this successfully. I suggest you try it out in a couple of accounts proactively.

Any information the customer gives you needs to be validated. Just because a decision maker asks about a feature doesn't mean it's essential to them. Salespeople who skip verifying what's actually important to the customer have lower closing rates. You might realize that something you positioned as a key selling point isn't that important!

When you listen to the customer's feedback and validate everything that they say, you give them confidence that you understand their process—or you're clearly trying to. If they see you veer off course, they'll be more likely to correct you because they *want* to give you their business. The product itself is probably only 20 or 30 percent of the value. The remaining 70 to 80 percent is your deployment process. You can't simply have the best product. You must also have the best strategy to put the product into place, with preferably lower, more manageable risk. That's what validating information allows you to do.

Why is it so critical? A few years ago, I worked on a deal with a bank in a small country. The CEO made it clear that he liked our offer, and he told us he would get his team to proceed. We thought we'd won. But here's the reality: A CEO is generally considered a Horizon 3 person. When any project goes live, it goes through the people on the ground. The Horizon 2 and Horizon 1 folks.. They're the ones who make it work. If you don't take the time to understand who wants what at every level of the organization and plan accordingly, you'll have a problem, no matter what the CEO tells you. We certainly did. We lost the deal because we didn't have the support of some frontline employees, and a competitor ambushed us. We thought the CEO was all-powerful. That's why Alignment (A in MAGIC) with Money, Authority and Need is a prerequisite. That's where politics come in. If you don't have everyone's support, you have no one's support; if you don't understand every stakeholder and reflect that back to them through validation, then don't assume you have a deal.

If you're worried you don't have the whole company's support, discuss it with your management team. Split the responsibility on the account coverage so others can help you get buy-in from the whole organization. When you go to this next level of making sure the customer understands you and you understand them, your win probability goes up to 75 to 90 percent.

Like clarifying small wins and minimizing risk, aligning payment schedules with outcomes also shows your customer you understand them. At that point, you don't have to play hardball. You won't feel like you have to *convince* the customer. You can simply look for their signals and validate their information through questions and answers.

What if you don't see any signals? That might be common in emerging technologies. Now, innovation is an iterative process. It's hard to predict your success ratio. Instead explain how you could lower the risks to his company. Clarity on the trade-offs will be critical in such cases. Expect purchases in small chunks. The "land and expand" is the way to go in such cases.

THE CUSTOMER-UNDERSTANDING CADENCE

Establishing a successful cadence in your day-to-day operations is crucial to overcome potential potholes, roadblocks, reckless drivers, and to get to your desired destination safely. This involves not only executing Act 1 with the customer and your internal stakeholders, but also with your management that you will call upon later for support. It is about setting expectations internally and gathering support for a worst-case scenario. Let's take pricing for instance. Pricing is a function of volume and competition. The customer wants your product at the lowest price, and your management wants to sell it at the highest. You simply want to ensure a win in a way it's acceptable for both parties. You start with a high number with customers, but a success cadence means starting with a low number with your management based on your early read of a qualified customer. Proactively feed information to your managers about outcomes, your product market fit, your product channel fit, and your competition. If you ask a customer for one hundred dollars and you expect to close at, say, sixty dollars, you have to pitch your management that this is likely a forty dollar deal. If you get serious resistance internally, then it's better to seek help to sell more value. Your managers or executives will be glad to contribute suggestions or to meet with the customers themselves. But if they like the construct of that deal for any reason, they'll support you well enough to enable you to play on your front foot with the customer. You will target a position when eventually the customer says, "I'll pay only, say, fifty dollars," and you'll get to ask if you meet that price, will they award the contract.

Most inexperienced salespeople leave the internal expectation setting till too late in the sales cycle and then struggle to get approvals. Or they simply tell themselves that the budget is too small and don't engage with the customer. Everyone in enterprise sales knows that different customers get different discounts for a variety of reasons. What justifies a 40 percent discount versus an 80 percent discount to two different customers? It's always the context. It's a silly mistake to think that the back-end approvers are being tough and unreasonable. On the contrary they are really trying to sell products at a higher value. Recognize that asking for a discount internally is really as important a negotiation as the one you have with your customer. Negotiations take time to help the other team understand the task at hand. The way around that is to spend sufficient time early in the process to allow for a full, internal negotiation cycle.

When I took up a new job there were many rules that I was not used to, like cutoff dates for approvals, cutoff dates for deal-booking of a certain value, and discount percentage approvals, and management happened to enforce these cutoffs hard that quarter. The following quarter, I had a couple of small deals that came in after the cutoff. I was mentally prepared to book them the following quarter, but to my pleasant surprise all the small deals got booked—not only my deals but everyone else's too. Why? Because the country numbers were short that quarter and needed to be increased.

It's often a matter of who up the management chain is doing well or not and if your deal can help salvage their situation. I have used this as my internal negotiation approach many times by being conservative regarding the dollar value and the close date of the deal. Customers also plan their purchases based on when you give them a lower price, like at the end of a quarter end or at year's end. Once I got into enough deals, I always made it a point to have hard negotiations at least a quarter out from what I thought was a closable date while pushing the customer to sign the contract in the current quarter. This allowed me to wait for my management to need more deals due to deals dropping in other regions. If I didn't, I wouldn't have to agree to the customer's price and could let the deal move to the next quarter. Every other quarter there was some deal slip or a loss, and all I had to do to sweeten the deal was move one of my next quarter deals to the current quarter. Everyone was happy, and I even got to play the "knight in shining armor" a few times. It is therefore your negotiation skill that leads you to create a win-win situation for all parties.

This approach works well in established companies, so you should leverage it to the fullest. In startups, however, my experience is different. While there is resistance, I have found the CEOs to be more willing to listen to their troops on the ground. Straight talk done early is good enough to get this done, but ensure that you spend enough time to allow for internal negotiations. The steps required to get this done, whether you are in a startup or an established company, are the same system. Look through the MAGIC lens of the Thought Plan and decide which approach fits best.

FOUR "TELLS" THAT SHOW THAT YOUR CUSTOMER KNOWS YOU UNDERSTAND THEM

How do you know you've successfully understood your customer and relayed that to them? I look for the customer to give me four "tells."

First, **they spend an increased amount of time with you.** Someone who's not interested won't keep making time to talk. These are busy people, just like you.

Second, **they demonstrate flexibility.** If they're showing that they're willing to work with you to find the best solution, that's a sure tell that they believe you're the person to get the job done.

Third, **they advise you.** If they're taking the time to tell you what can be done and what should not be done, that means they're rooting for you. When the customer shows you where the money comes from, who's in charge, what they're looking for, and the best way to get approvals, you know you understand that customer better than anybody. I saw this tell very clearly with a customer whose budget we overshot. Their budget was $300,000, but our proposal was over $700,000. Instead of going straight into negotiations, we asked to talk to other decision makers within the company whose success also depended on the project. They guided us to win the deal.

And fourth, **you have a champion at your customer's company**. Most salespeople build their own "champions," meaning contacts within the prospect's organization who support the deal. Your champion is trying to get a solution deployed. They're self-incentivized to move the sales process along; your product or service would make their life so much easier, and they simply cannot wait to use it.

Unfortunately, there are also antichampions. These people just don't like you, or simply prefer another vendor. If your champion is convinced you understand their needs, they'll give you access to the decision makers you must talk to. Your champion can present your offer concisely to these individuals in the internal discussions you're not privy to.

You can have any number of champions in an organization trying to sell you to their decision makers. I've learned that having more than one champion is an indication that you really understand the customer. When you ask questions and educate them on how your solution aligns with their top priorities, your number of champions increases. Your "intel" improves.

You don't have to be the smartest person around to see these four tells. And when you see them, show the evidence to your management to build up internal support for your deal.

By internal support, I also mean support from your partners. Partnerships where multiple companies collaborate on a proposal are common in enterprise sales. Partners often have a relationship with the customer from previous contracts or from their current service. In those situations, coach your partner. Guide them into the same understanding that you have so they can show the customer they also understand them. This is much smarter than doing all the work yourself simply because of the time and effort you'll save. It becomes a virtuous cycle in which you coach each other to win the deal together.

As a result of creating win-wins with partners, you will expand your circle of influence. I've had partners come to me asking how to handle a situation with their customers. I've become a source of intelligence for these partners, and they've shared their own customers' project details with me, which has led me to find ways to sell my own company's products or services into those accounts. This is how you take control of your own career. Be the source of knowledge for others, connect the dots for your partners, and you will have a higher trajectory. You get to enjoy the selling process, strengthen yourself, and achieve your career goals.

As a salesperson who understands the big picture, you'll also get opportunities to fix broken territories. I was not a fixer when my career started, but I realized that when you demonstrate skill in understanding customers (and teaching that understanding to others), underperforming territories come to you. That's how management functions. If you do well, you receive more responsibilities (and more earning potential). Add value, and you receive value right back.

THREE MAGIC STEPS TO ENSURE YOUR CUSTOMER KNOWS YOU UNDERSTAND THEM

Ensuring customers know that you understand them better than your competition is really the magic sauce. This is something you will notice every successful salesperson does either consciously or simply based on instinct. It doesn't mean understanding the full need perfectly. It's just that you understand it better than your competition and have approached the rest of the sales cycle based on that understanding. You can achieve this with a combination of:

1. **Using your own product knowledge.** This includes details of how other customers are using your products in a measurable way. What's the alternate way people address this issue without using your product? You can do this preparation yourself within your company most of the time. Don't approach this like a chore. Don't memorize a list of qualification questions. This is the easiest part, so don't score less than 100 percent.
2. **Treating the prospective customer as unique.** Have you heard the saying "What got you here won't get you there?" What gets you to the table is the commonality between the customer's industry or function and your company's reputation, but what gets you to win is the uniqueness of that customer and how well you are aligned with that. Try to know everything about the "hole" the customer is planning.
3. **Convincing them or being convinced.** Always aim to be frictionless. It was true before. This will be a way of life in the post-COVID world. Physical boundaries don't matter if all customers get used to Zoom calls. You will have an unfair advantage if you can make working with you frictionless. Keep these four magic words top of mind: Convince or be convinced.

If you need to make sure you understand your customer better than your competitor does, then how can you know how much your competitor understands? You can only see it based on the customer's behavior.

Remember the telecom group that was insistent it needed a data model that we turned around? Here's how that deal maps onto the three-step process.

1. **Use your own product knowledge.** We knew that the company's enterprise data warehouse (EDW) had a data model at its core, but we didn't have a solution, whereas our competition did. We also knew why some customers addressed the data model requirement in another way, so we peeled back the layers of how they were addressing their data model before making a decision and asked if they were unhappy with it.

(See diagram in "Yeah. . . But" section in Act 1.)

2. **Treat the prospective customer as unique.** Since we knew we didn't have a data model, we asked how many KPIs, dimensions, and reports that customer was currently generating and if they were happy with them. We also asked how many more reports they expected to generate in the coming years? That understanding early on ensured that we didn't have to partner with any data model providers to present a full solution. Had we found the client was unhappy with their current solution, we would have taken a different approach. Remember, every customer is unique!
3. **Convince or be convinced.** In order to execute this well your approach needs to be rooted in first principles. Start by identifying assumptions. Then we asked powerful questions to break the problem and came to the point. And create a solution from ground up. This could have easily gone in other ways. Getting to the root of the customer's needs and knowing your own technology/product cold will help you stay objective in your decision-making.

Follow the three-step process, and customers will tend to give you signals that you're the one for them. So, how did we know our plan worked?

In a future conversation, we touched on the data model with the EDW team. "We heard Top Brand pitched you a data model solution, effectively saying that all the efforts your team has put in these past years is a waste."

"Yes," the EDW head replied, "we did get that proposition. But we're not considering it."

"It's interesting that they didn't bother asking whether you needed it or not," I added. "Sounds like they simply pitched what they had in their bag."

"Yes, they talk more than they listen."

Validate the fact that you understand your customer, and use that to differentiate you from your competition. Keep talking to customers and use the MAGIC lens to measure your situation accurately.

Here's another example of using the three magic steps.

I was an individual contributor for a regional news media company (a different one than the previous example). I had spoken with their CFO for a database sale on the phone, but I had never met him during our twelve-month engagement. We closed a deal at the end of November. About eight months prior, we found out they were looking to set up a knowledge portal (i.e., middleware). They were evaluating a market leader and another 2 vendors. At that time, my company wasn't known for middleware, even though just a few months before, we had launched a major release of middleware. No one in APAC has bought the new release yet. Keeping in line with my approach, I read up on all the new features of the product, and validated it by seeing if customers valued those features.

I went back to the CFO in early December and told him we had a proposition to help him save money on his knowledge portal.

"What is it?" he asked. "You guys don't have a solution."

"You're partly right, sir," I said. "We didn't have it before, but we've launched a new major release that has all the necessary features you're looking for. I'm aware you're in the final stages of negotiations with the Big Boy leader. And December is the end of the year, so they'll likely give you a good price for closing now. Your team will either like Big Boy's solution or they won't, but if you evaluate us as well, we'll move at double time to complete all the steps with your technical team in line with your timeline so you can still make a decision by the end of December. If your team likes our solution, you'll definitely save more, as our price is lower. And if they don't, evaluating us will get you an even lower price from Big Boy because they won't want to lose your sale."

"Yes, I understand your offer, thanks," the CFO said. "But why do you want to bid for a deal that you have such a slim chance of winning? Why do you want to push your team so hard just to give me a price advantage?"

"I won't spend my energy on a deal if I can't see a chance to win," I said, "let alone waste my team's efforts. But when I went through our new release features, I actually thought that your team would like our solution. I wouldn't be surprised if they selected our product instead of Big Boy's. Even though it's a new product and therefore 'unproven,' I can guarantee that we will do whatever it takes to make it successful for early adopters. I'll share my plan on how we'll help you succeed with it if your team recommends us."

In line with the three-step process, before I decided to make this bold pitch, I went through the details of the product, I saw where it fit, I looked at what situations customers used it for, and I researched our competitor's offer. That preparation helped me build my credibility and get a seat at the table. Better than sitting back and letting Big Boy win.

I aligned my intel with the customer's unique situation. I didn't just give them a general pitch. I was honest with them and gave a very rational reason as to why I wouldn't be wasting my time if there was no real chance of winning. I also explained that there was no downside for the customer to give us a chance. That opened a magic door, like an antenna positioned to catch all incoming signals.

Also notice the frictionless way that my conversation created compelling reasons for the company to give us a chance. What did the CFO have to lose? The only way they could have said no was if they'd already made up their mind. My bet was that if they said yes to the evaluation, we had a 50 percent chance of winning the deal. As it turned out, it was wide open for us, despite there being no scope, but that's a different point I'll share later!

Since the evaluation criteria were all ready, the CFO had to share them. It included one or two presentations and a proof of concept (POC). We agreed that two weeks would be a good timeframe for us to complete everything. The CFO also sent a team of people to our office for a product deep dive.

Luckily for us, December was quiet on the deal front, so I could line up all the right people for this sale. While my agreement with the customer was two weeks, I told my team that they should work as if we only had seven days. I made sure I was on standby to help coordinate any critical activities with the customer. I wanted to wow the customer with our agility and timeliness. I shared this game plan with my team and provided clarity on what role everyone had to play. This was essential because, without context, the technical team could have thought the sale was unreasonable.

The team did a great job, and we completed the POC in eight days. After a full review to make sure we didn't miss anything, we went back to the CFO on day ten with the POC.

He was surprised. "Are you sure you did a good job on it? I was sort of expecting you to ask for an extension on the timeline. Big Boy took ten weeks to do it. How can you do it in less than two?"

I said, "I thought you might say that, sir. Yes, the POC is complete and at a high standard. We're fortunate to have a very skilled team working on this deal."

Understanding the customer is a continuous, two-way street in every deal. If you think you need to understand the customer and you leave the validation part out, you're just spraying and praying. When we proved that we could complete a POC in less than two weeks while our competitor took ten, we got a seat at the negotiation table.

To minimize our risk of losing, I thought it might be a good idea to delay the procurement until after December, since Big Boy was likely to give higher discounts at the end of the year. Salespeople tend to track deals with intensity at the end of a year versus in the first month of a new fiscal year. My next step, therefore, was to delay their decision until at least January. We had our product release on Linux first. Followed by other operating systems. The problem was our customer standardized on another operating system. We delayed so that we could go through the process of how to adopt Linux, why it was a good platform, and organized some training and planning for the customer. Two months later we closed that deal, which was a first in APAC for us. It took us four months longer and a disproportionate amount of effort than the planned implementation time to get the product to work.

We implemented the three magic steps three times within this short sales cycle. You'll probably use it dozens of times in most deals. Why is this important? Because it keeps the sales team alert and objective. First of all, it ensures you're well versed on your products to better identify opportunities. Second, if the sales team gives a presentation on a certain product a few times, it's easy to assume that you know the product in and out and complacency sets in. By following a constant validation process throughout the sales cycle, the red flags show up quicker, giving you enough time to make adjustments. And third, it simply improves your probability of winning or, at the very least, leaves the ball in your court. That's a fantastic position to be in.

"The weakest link in a chain determines the strength of the chain." This is 100 percent true in sales. With the regional media company, we had to get a seat at the table by Aligning to the outcome and growing our influence, then cover the MAN in the Context of the deal. Even if one of them was weak, we would've wasted our time, not to mention the negative impact it would have had on team morale. It is your responsibility as a sales leader to execute the MAGIC process effectively.

The MAGIC lens provides you with the clarity of thought to delegate, decide, and motivate teams; it gives you the ability to think on your feet and negotiate easily. You will magically find yourself a lot more competent in quarterbacking any deal.

One large global US conglomerate set up several subsidiaries in India. They didn't buy our software locally.

During a phone call, one of the sales directors told me that even if we didn't get a deal locally, it was our responsibility to continue to engage with them to prevent any competitors from replacing our footprint. It also ensured that the customers were happy. It was a nice "social good" approach, but you can only expect what you can inspect. If there was no measurement for success, how would anyone know if we were doing a good job or not? So I followed an approach that later became so successful, our SVP for APAC asked me to write a memo on how to sell into large global accounts and grow them in APAC.

The lynchpin of this approach is, of course, to ensure we understand customers better than others. We had the solution part ready, thanks to the work done by my predecessors, but how did we ensure these GE subsidiaries knew we understood them better than our competition?

I started with the procurement teams in each subsidiary and introduced myself as a single point of contact. Their KPIs were easy to guess. If we gave them better terms for the same product, they'd tell us who the buyers were. Part A of the plan was to get something on the table. What can our local procurement teams do when decisions are happening at HQ in the US? If I could improve the offer (and I wasn't referring to additional discounts), I could get them interested. The procurement guys introduced us to MAN, and we started doing product update webinars as a way to keep them updated on our technology. No asks, just sharing information. This was executed with local partners who also wanted access to try selling their services.

We planned the next set of webinars for each individual subsidiary instead of inviting every team to all of them. In preparation, we asked each subsidiary if they'd like us to cover solutions to any problems they'd been having. They shared their expansion plans for their new and existing projects. This let us see what additional products would help their operations. It also showed us how they shared their requirements with their HQ.

Once we knew the requirements, we restated all their needs to them. We told them if they helped us get a deal locally, we would still give them a special global price and also offer some "free" service for a few days via partners. It was simple to make their procurement teams look like winners because they were getting a better deal than what was negotiated with HQ. There had already been talk of trying to delegate purchasing locally to their in-country teams. I didn't know HQ's plans for a fact, but we either created them or accelerated the process.

One of the most powerful customer questions I learned from our retail business unit leader Aylwin was, "Are you happy?" A customer may be using a product, either yours or another vendor's. If we intend to upsell, cross-sell or, if it's a competitor account, replace the current product, the best way forward is to find out if they're happy with it. Nobody wants to admit they made a big mistake with a previous purchase, so you need to be delicate. Asking smart, sensible questions without sounding like a reporter is a skill you need to develop. My advice is to ensure you have as many prospects as possible and have the motto

"Keep talking to customers" in your head all the time. Experiment with your questioning and never treat your customers like just part of a segment.

For instance, I was working on a deal with a fast-growing retailer. They needed a couple of enterprise resource planning (ERP) modules. Five years earlier, they had deployed a merchandising solution that offered basic ERP functionalities. Now that the company had grown, they needed a bit more sophistication, so they invited us and the market leader to pitch a solution.

The retailer was in the market leader's territory, with us as a distant second. So how did Aylwin turn this around? He reached out to the retailer's CEO. Armed with knowledge of their unique situation and of the retail industry, Aylwin asked a direct question: "Are you just going to evaluate ERP modules and compare features, or would you prefer a solution that lowers your inventory costs and also solves your ERP module problem?"

"Well, of course I'd prefer to lower inventory costs," the CEO replied. "That's the goal."

Retail is heavy on inventory costs. Even a small percentage reduction translates into big savings dollars. From that question alone, Frank knew we understood them better than the competition. That allowed us to lead the conversation in a totally different direction (in what was a relatively weak ecosystem for us), generating a 40 percent premium and several times the value of the budget the retailer had originally planned.

Context is everything. Not all deals will be as elaborate as the ones I've mentioned. You may have more transactional sales with shorter sales cycles, large enterprise agreements, complex transformational deals, and innovation deals that are a bit hypothetical. Innovation is an umbrella topic where enterprises work better with their customers, partners, and vendors to survive, adapt, and thrive in this disruptive world.

CUSTOMER READINESS

One of the most important things to pay attention to in sales is the readiness of your customer. In an ideal world, a customer's readiness to evaluate, purchase, have a budget ready, and so on, is always easy to interpret. In reality, you'll find a constant disconnect in practically every company. A lack of understanding of customer readiness shows up in a pipeline requirement of 4x or 5x, which is just covering up the real problem. Why do you think forecast accuracy is important? How much deviation is allowed, and what is your track record on your own forecast accuracy? A huge amount of time is wasted across organizations to manage this time-wasting work.

Call it a "cleaning pipe," validating close dates, or whatever you prefer, but I think sales teams can improve their productivity by 30 to 50 percent if this process is followed. The results tend to show as soon as you execute these in your opportunities and deals. I noticed that most of my peers worked on 4x or even 5x pipelines compared to my pipeline average of 1.2x to 1.5x. In my mind, any such situation means the status quo must change. Time to deep dive and find out the reality in the pipe. It's no different from having a multilayered bureaucratic or red tape process. What's worse? The wrong diagnosis! This leads to the wrong fixes, which lead to more cover-up stories, which lead to politics in office, which lead to tough performance reviews, lower commissions, and higher attrition rates. You get the picture. All of this will happen one way or another even if they don't happen at the same time—that is, unless you increase the accuracy of your sales pipeline .

If the problem is that of accuracy around your customer readiness assessment and how to reflect that in your CRM, then this is really a skills issue that can be fixed with sales coaching/training. Any rep involved properly in a sales process will know at the beginning of the quarter if a deal is likely to close or not. Some inexperience or low-skilled reps may have "happy ears". I'm not suggesting there won't be fluctuations in the pipeline or that you need a cushion. What I'm saying is that willfully (or because of lack of skill) maintaining a 4x to 5x pipeline with the wrong close dates hides the real issues.

Every time I was given the opportunity to turn around underperforming teams or territories, this is what I fixed.

How quickly can this be fixed? The turnaround from fixing the pipeline to better results shows up within three months in my experience. This is across solutions, countries, and cultures and in big and small companies. In a midsize company we closed thirty logos in fifteen months while hiring, building partners, and selling. Startups want those results to improve valuations.

We visited and reviewed every deal in the pipeline in the region and closed two thirds of them. Though initially disappointing, it freed up resources to be deployed in account with customers who scored high on readiness to buy.

I have learned this hard lesson on readiness even in home situations with my kids. In my eagerness to help them grow strong and independent, I have often tried to share thoughts or discussions that were simply too difficult for them to comprehend or not in line with their experience. These lessons ended up being misunderstood, not to mention time wasters, as you have to do the cleanup. What a waste.

CREATING AN OPPORTUNITY VERSUS INTERCEPTING AN OPPORTUNITY

You need to absolutely pay attention to whether you are creating an opportunity or intercepting and responding to an opportunity your customer or another vendor created. Intercepting an opportunity is when the customer intends to fix an identified problem, already did some work, and has invited you to present as a part of the evaluation. Secondly, you may start working with customers to help them realize a problem they haven't identified or prioritized, or you may get involved in the middle of a sales cycle and try to influence and persuade the customer to move any specifications to your company along the way. Be open to the fact that three different customers in the same country, even in the same industry, may be in three different categories, even though your product or solution is the same. If you proactively look for uniqueness instead of similarities with the aim of repeatability, you will be very successful. You need to use the similarities only to get to the table. Once you are at the table, stop looking for similarities. You should be worried if there are too many similarities. Either you are not looking hard enough or it's a risky deal to win. Just watch out. You need to understand it's not that difficult to find the uniqueness once you do it a couple of times. Some accounts will be harder than others but the trade-off is higher win rates, less waste of time and resources, and better career/monetary growth.

I'll share two painful instances where we didn't follow these guidelines and faced negative consequences.

HOW TO FAIL AT UNDERSTANDING CUSTOMERS

I'd like to tell you about a data lake deal. There was a customer hostile toward our company due to bad relationships that had been fostered over negative sales practices. As a result, there were senior officials who didn't want to invite us for the bid, but the company's procurement team sent the bid to multiple vendors, including us.

This is a clear interception deal. We didn't do anything to create this opportunity, and we were starting negatively. There were no local partners willing to work with us, but there were multiple partners willing to work with our competitor. What's more, my team, which consisted of big data and analytics, didn't have any prior relationships with this account. We decided to participate since our team was just forming and didn't have much pipeline, despite the low odds. We assembled a regional team that did not include anyone who had fallen out of the customer's good graces. The first part went out smoothly based on our read of the RFP, our elaborate preparation, practiced role plays, and several dress rehearsals. If we didn't stand out head and shoulders from the crowded vendor list, there was no chance of winning this. Even a narrow win was not possible, as some customers didn't want us to win.

The first session onward was perfect and well above the customer's expectations. They liked our proposal because it was a comprehensive response, and when we presented, there was a team of technical evaluators who were real-time validating our claims on features, functionalities, and their fit to the process. The People, Process and Technology part was hit out of the park, and the customer's tone changed. The people who were coming close to "blows" with our company, threatening legal action, started sending feelers that they liked our solution the best and that they are keen and open to work with us.

That lead lasted for about three months, and all our background checks revealed that the customer was convinced that our team understood their needs the best. It was a full stack on software and hardware and services. Our services were regional and, in general, considered expensive. We didn't usually position our services unless we saw a big risk of something going wrong with implementation or if it was a strong account of ours where the risk of losing the deal was really low due to our existing customer intel and relationships and understanding.

In this case, however, we positioned our services directly because no partner was willing to work with us on that deal. (We had approached four before submitting a direct bid. They knew about the bad relationship and didn't give us any chance to win.) We planned to handle the objections (look for Objection handling techniques in chapter 6. This one falls under the 'genuine problem' category).

One thing you should know is that in Thailand, where we were pitching this deal, it's common knowledge that it's better to have a local partner to help with selling, supporting, and implementing the deal language. The issue started there. We were now too deep into the sales cycle to find a local partner, and we were nearing the price negotiations. One question that came up from a key evaluator was if we were open to mixing and matching our solution, as they liked some parts of our proposal and some parts of our competition's proposal. That was when I made a mistake. Instead of asking which part of the proposal they liked and its rationale, I recollect saying something that sounded like we were not open to this or reluctant to be open. I was thinking a lower bill of material would not be able to get the same high discount internally at our company, and I didn't know if we could work with a direct competitor. Even as I said that I felt something not right about the way I had answered, but I went ahead anyway.

It wasn't a mistake. Blunder is a better description. Don't expect Thai customers to request the same thing again if you reject them once or even show reluctance. Customer never asked me this question again, so I wasted a wonderful opportunity to get deep into the "uniqueness" of the customer. The deal cracked on this "uniqueness" and resulted in me breaking my own rule of "convince or be convinced." A month later, we ended up being a check in the box of procurement's evaluation and lost the deal to our competitor. It was a thorough disappointment even though we knew the odds of winning this deal were very low to start with. It was within our grasp, and I was stupid to not strike sensibly. What I realized later, to my shock, was that the proposal that customers liked to go with was our end-to-end footprint. The thing that they wanted to change was our services. They were more comfortable with a local partner (who bid for competition and therefore directly competed against us) who was more "flexible" and "familiar" to work with. Alas, I thought they wanted to mix and match some products! This is why they say what they do about assumptions.

All I needed was to speak for a few minutes extra to understand what the evaluator was asking. It troubles me every time I think about it.

This was a lesson I learned so well that when a large bank, another high-profile customer, similarly stated that they preferred their current services partner (Global SI) to our Consulting Services as they liked Global SI's SLAs and the skills. Remarkably, the Global SI has quoted our competition in that bid. I was on a phone call with the bank's decision maker when they asked if we are open to a mix and match. I asked them what it was they wanted to mix and match. They said services. My next question was if they would release the PO at the quoted rate if we allowed the change. I confirmed in two seconds that it would be done and that I'd handle internal approvals. (These things are never easy internally either, but that's a different story I'll discuss in this book.) We won that deal.

It's amazing how things show up like magic when we do these basics right.

Yet another big lesson learned was in the another Bank deal. We were ahead of all our competitors. The situation changed, though, after seven months of fully involved effort. Our validations were consistent in proving we were on track. This was also a digital transformation initiative. The evaluation team was from the bank's regular team, and we knew they would hire someone to run the platform operations for this project. We discussed pricing, knowing the ballpark budgets and preparing well with partners on how to close the deal quickly. One week before the bank was to award the project they onboarded a new person—we learned of this hiring much later, after it was too late—who happened to work on a platform that was not even a part of the evaluation process. All of a sudden, the Bank asked us to compare our product with that third vendor, and we reacted by sharing some of our product's features. We had no clue that the person who was hired recommended that third product simply because he used to work on the platform and not on ours.

You see, having last-minute competition is not new in sales. There are ways to win by being a last-minute entrant, and there are ways to defend against any last-minute entrants. If we follow the process outlined in this book fully, it works like an immunity booster and defends against last-minute surprises. Had we used that opportunity to ask the customer why they wanted to compare us to a new vendor that late in the sales cycle and spent some time with them instead of responding with comparison documents, we wouldn't have ceded control to the competition and played their game. Instead, we were legitimizing their seat at the table as an equal contender. That knowledge of a new hire would have guided us to address the objections better, to reestablish our credentials as *the* vendor who understood the Bank the best, and to try to address the risk perception that our platform was not something that the new hire knew. Additionally, we would have taken precautions to learn whether the new competitor didn't do other things we did and to highlight those risks to rally some support within the Union Bank team. The fact is the rest of the Bank team didn't know much about that new vendor as they didn't perform a full evaluation. We would also have had the option of reconsidering our pricing strategy. We had enough champions in the bank to get the necessary guidance, but didn't even know we had to use them. Since we assumed this was a checklist comparison, it diluted a customer's knowledge on who understood them better among three vendors. 3 years after we lost that deal, we replaced that product with our product because the Bank eventually realized that it was a wrong choice and we won it back.

In emerging technology products where there is no clear market leader and truly matured skill sets are scarce, you need to be extra careful. We need to double down on this process as your "bread and butter" to stay the course and prevail.

Luckily these losses were rare and approximately 90 percent of the time I was on the winning side when we competed.

In another case, it was our turn to make a last-minute entry against a core banking system that mostly ran on competition databases, with none on our database . The partner who proposed this deal was owned by the CIO of that bank. The core banking vendor also had a majority of their deployments on competition databases. We won that deal with the combination of our product knowledge, our fit with the customer's uniqueness, and a frictionless approach.

STOP STOPPING YOUR WINS, START STARTING MORE WINS

Let me share with you some comments I have heard over the years that gave me a deeper perspective. Except for two the rest came from folks who were not in my team. They were either my peers, my managers, or even my customers.

1. After a customer meeting with me in Thailand, my new manager at the time said, "I never knew the product could be positioned that way. Now I know why you get those results."
2. Susan, a Sales Director on another team, once said, "Yes, I know. You disappear, and eight or nine months later, you return with some mega deals, like magic out of nowhere."
3. Our Senior Vice President said, during my quarterly review, "You don't have offices in your territory, not enough skill sets on the ground, and the economies you are managing are third-world economies that everyone considers difficult. Then how are you able to pull off this magic with our latest product?"
4. A colleague I had never met before ran into me at a lobby while waiting for an elevator and said, "I heard you bring out the best in people. Can I work for you?"
5. Our VP of channels told a US company's APAC head that if I were convinced by their solution, then the magic would happen.
6. Joseph said, "Boss, I didn't think you'd come out of that meeting alive. When we brought you into the meeting, it had 'disaster' written all over it. But the way you dribbled in several of the gotcha questions was magical. How did you do that?"
7. My manager told a partner, "If you want creative ideas, you should talk to Raju. He is full of new ideas and an unusual perspective."

8. Another manager addressed our team of twenty-five, saying, "I don't know how you guys pull off such results. My job is to just approve whatever you ask for. Keep it up!"
9. Tawatchai once said, "That customer never agreed to buy from me after repeated tries, and you met him once and you moved his fifth priority (our product) to first priority by the time we finished this meeting, in *front* of me! How did you do it??"
10. Our customer's CIO told our competitor that he didn't understand why they approached him for projects. "Why can't you work like *our company*? They come in, create a project, and take the money!"

There are many other instances and I'm not going to list all of them, but what you see as different ways of creating sales are nothing but a simple, reusable process that anyone can learn irrespective of the state or stage of their career. You can create your own flavor of magic. It will both be unique and effective.

Whenever I go to a territory that hasn't performed well and my job is to fix it, I ask a simple, straightforward question of what's going on with their performance and what they think should be done to improve it. I encourage accuracy over political correctness.

The answers loosely add up to how guys coming from regional HQ don't connect with the local culture, how we are perceived as aggressive, just given sales quotas but no sales support except for chasing numbers or a variation of that. Interestingly no one has ever said they didn't have the skill or that they weren't suited for the job.

I used to have a standard approach. "OK, I hear you. Here is my proposal. . . You get me the meetings with your toughest prospects and customers (who have budgets to spend, of course) that don't want to meet you, or don't prefer to work with you, or don't like our company and prefer our competitors. Since you know the lay of the land, I'm sure you will open the doors for me. I won't bore you with all the good things I have done in other countries, customers, as they are not relevant to your context in this country, this vertical, and this account for this situation. Here is the deal. I'll help you sell to them right in front of you.

If I succeed, I want you to repeat it with other customers. If I can't sell, then I won't chase you to close such deals, but there is a third condition. If I sell to your toughest customers and you can't repeat it, then I think you are the problem and you have to resign. I will not look to fire you or find your replacement until we reach those conditions. Deal?"

The thing with magical selling is that it is based on First Principles (not analogies). It creates each piece that you need as an ingredient for magic to happen and is naturally aligned to the next piece as if they were magnetic. They just fit in well. The problem with different selling methodologies or designations like "consultative selling," creative selling, hunter selling, farmer selling, and others are really the byproducts of adapting this approach. This is why they may appear theoretical to salespeople. You tend to identify only with part of those processes and therefore it is easy to forget the entirety of them a short while later. The reason why those naming conventions exist is because there are countless scenarios out there. They are like the hats you wear depending on the situation with your current sale in your specific customer at that specific time.

Let me try to simplify it further. All those names you hear about or give to a selling style are like the descriptions of flowers or fruit. They are all relevant in different circumstances, but if you think the way to make flowers and/or fruits is the ingredients—soil, water, fertilizer, sunlight, and seeds—that results in the byproducts that are flowers and fruit. Does it sound possible? It happens every day around the world.

This approach helps in selling in a variety of scenarios and also in hiring. But does it help with your interviews if you are applying for a role? Let's see. . .

A friend of mine who I mentored for a while, had her role made redundant during a merger. At that time she was an individual contributor running sales in some emerging countries. She was a solid performer and got good results, but she had no prior management experience in any formal role. For about twelve months after she lost her job, she landed a couple of gigs that didn't last long—first as a first employee in APAC for a US company where she lasted for six months and then as a manager of a team of some eight or nine people at a large company for a month before deciding to quit again.

I asked her why she was doing business as usual in her job search, and we discussed options on how to approach the hunt. I suggested that she follow the approach of doing some groundwork and pitched it as understanding the customer (the hiring manager, in this case), informing them about where their revenue is (i.e., do homework first and have a directionally accurate business plan) and how she could help bring that revenue in and share the skills she possessed that fit into that role. She reached out to the CEO of a company in the United States via LinkedIn and proposed to speak with them. There was no headhunter calling or job description to respond to by customizing her CV. Mind you, she didn't have any real people management role prior to that. Neither did she have experience selling into all the countries in APAC. It's hard to imagine for people how a profile like that convinced the CEO of a US company to make her the person in charge of Japan and Asia Pacific (JAPAC), but three months after she started that process, she got selected as their head of JAPAC business. We had several calls during the course of that interview process where I was simply guiding her through this magical selling approach. After all, she was selling her skills and trying to persuade the CEO and his management team. I didn't call it magical selling at that time or tell her the framework to follow blindly. It was a step-by-step process of guiding her to the next level, and she pulled it off. All the work was done by her. She was the same person with the same experience and same skills, but approached it with the magical ingredients and took that spot. Don't assume it is impossible.

She asked me to recommend some candidates to hire to her team. I spoke to someone in Australia who happened to know her. When I told him that she was now heading JAPAC for this company, he couldn't believe his ears. He said, "We worked in the same territory for different teams. I know her. How on earth did she get a JAPAC role? It's amazing." He found it unbelievable, but she knew exactly what happened and how it happened. It's not that the whole process went on without any back and forth. Many times she realized this approach I described works even though she didn't believe it when I said it. Every time it got her a result, she came back with something like, "You are always right." I like the sound of that, but the point is not whether I'm right or wrong. The point is really if you know what's required to propel your career to the next level quickly. I have no doubt you will start seeing results you haven't seen before!

One such success is a data point. Two in a row is a pattern. Within a year after she got that JAPAC job another individual contributor got selected to run JAPAC for an Insurance solution provider. Same approach.

Yet another candidate from a technical pre-sales position used the M in MAGIC (Money, Authority and Need concept) to land a sales role he was eyeing for a long time. He sounded ecstatic when he got his dream role. I know it sounds like magic. So try this process. What do you have to lose? Stop stopping your growth prospects. Start a new career path to the next orbit.

This may seem like a counterpoint to some, but when I do forecast calls or plan a new year or new product GTM, I look for what is different and unique. It seemed odd that all companies tend to look at similarities and categorize them into segments. That reduces efficiency. If you need 4x to 5x qualified pipe, I see it as 20 to 25 percent efficiency. I don't like those odds. In other words, it means you don't have control on your territory, not qualifying well and not having the right skill to do so. Remember, the day you qualify a deal and mark it as a real opportunity, you need to have put enough thought into how to make it winnable. With my team, I always try to check their effectiveness by finding out how unique they think (or know) each of their opportunities are. Any number of times in forecast calls if I hear that a certain person says a deal will close a certain date (as opposed to the number of signatures required, their whereabouts, if they know this deal is coming for approval, and all their objections have been taken care of), I call out that deal in the forecast as at risk. **If you are constantly facing new situations in every deal, you are increasing your growth potential all the time.**

Look for uniqueness in every deal, no matter how many similarities they may have at first glance. Those who use the knowledge of those similarities *and* the knowledge of uniqueness end up creating winning angles for themselves. Somehow they never seem to need 4x to 5x pipelines.

ACT 3

CHAPTER 4

Build Partnerships (with Employees, Customers, Resellers, and Everyone Else)

Treating partnerships as different silos from your organization is a terrible mistake. It's amazing how many companies make that mistake routinely and just do lip service to the partnership or leave it up to the partnering manager to make the collaboration work.

Let's start at the beginning. There is a market need to solve a certain problem that a company recognizes and develops a product or service in a way customers see a significant benefit. Now the company needs to find a way to serve every organization who needs that solution, to support them, service them, and grow their business. Then they need to go about hiring sales teams in different territories with relevant experience, the necessary network, and so on so that they can identify a handful of potential accounts. Unknowingly they have started a bottleneck. Why? Because first you can never hire a sales team that will reach the full market, and second, even if they could, they can't reach that market fast enough. If time is money, then in my view, not having a partnership plan from the starting gate is one of the worst decisions you can make. The Go-to-Market (GTM) strategy needs to proactively define and identify the best way to build channels and alliances to reach the customers we plan to serve. "Proactively" is the keyword here. Some reps think they can maximize revenue better if they don't share with partners. This is true in some cases, but if you check across your territory and even some large multinational accounts, there are a lot more cases in which partners can help you maximize. Context is therefore important.

For instance, we spent nearly a year with a large, hostile account to turn it around. No partner helped us, and in fact they were selling the competition's products during that time. We finally turned it around and started winning some deals and felt good about our control on those opportunities. The team was motivated, got the alignment, and things were looking up.

One of the active partners then approached my sales rep and offered to work together. My rep didn't think we could trust the partner but arranged a meeting between the partner and me anyway. I mentioned to the partner that we were winning some deals and meeting our revenue targets without the help of that partner, so why should we consider working with them? If they could give proof that they can help us win more, I was open to partnering, but I wasn't looking to just let them take the margin on deals we are already closing. The partner said that they knew some of the deals we were closing, but we were not covering other deals happening in the same account that they knew of. I assured them that if they shared the details of those other deals and we didn't know about them, then I would let them lead those opportunities immediately. The partner shared two new projects that we didn't know about, which were led by the competition and that the partner could help us win. As you may have guessed, I agreed to let them lead and we bagged $700,000 of the competition's lunch.

The moral of the story? Even in your stronghold accounts, don't assume you have everything covered. Have a direct engagement to protect your turf and also leverage your partners' strengths to Grow your Influence (G in MAGIC)

If you establish yourself with some wins while you are convincing partners to work with you, you are setting yourself to successfully scale your revenue. My personal rule of thumb is that 30 percent of your addressable market needs to be sales-rep led and 70 percent partner led. (Note that sales can still be present on these accounts but in more of a support role.)

So, why can't so many sales reps and managers see what is so obvious? One reason I noticed is a myopic view that results in a lack of readiness, but a more common reason in my experience is that they lack the skill to build partnerships effectively. They've never done it before, have formed opinions with superficial information, and then refuse to learn if there is a better way.

While selecting what kind of partners to pursue, your number-one should be alignment. Call it Product-Channel Fit. Just like Product-Market fit. There are dozens of partnership models out there that can be confusing. Define the one that aligns your product to the channel. It should not look like you are selling cars through a pharmacy. That's an exaggeration, but you get the point. So, how do we decide when and which partner to work with? Let's see if our MAGIC lens helps us. Consider the following five factors on a scale of one to five. Ask yourself these questions in relation to potential partners.

M: Are we covering Money, Authority, and Need (MAN)? Do we know what their Horizon 1, 2, and 3 outcomes are? If not, who can help me internally? And who can help me externally? Who are those internal and external partners?

A: Are we Aligned? Can we prove that we can deliver their success measurements or can set success metrics (if the customer doesn't have the maturity to do it themselves)? Who will do this? Are there gaps in doing it well? Do we need a partner to strengthen our alignment?

G: Let's talk Growth. What's the current status of influence? Who's competing with us in the account? Can we do this alone? If not, who we need to move to our side and in what way?

I: Next, Inspect what parallel activities you are running. What's your cadence? What are your roles and responsibilities?

C: Do we understand the Context fully? Are we speaking the MAN's language? Have we done this type of deal before? Is it the best use of my time to do all the paperwork and groundwork or would I have better context with a partner? What strengths should the partner have? Why would they work with you?

If any of these rank less than four out of five, it's time to bring in partners. Personally I'd like to use partners even if we have all fours on an account because it would free my time to target other customers or other activities—the difficult and strategic ones like hiring, for example, that need more attention.

The Asia Pacific market is built on partnerships between companies. In enterprise sales, I haven't seen companies scale well if they don't treat partner ecosystem development as important in the way that they do customer acquisition. Maybe we should call it partner acquisition. This is regardless of whether you run a large company, introduce a new product, enter new countries, or lead transformation. Early in my career, I learned about measuring the ecosystem's value. For every dollar of licensing a certain product, the total ecosystem value is ten dollars. That includes revenue from product support, maintenance, services, training, implementation, and other value-adds. I don't know if that's a scientific study or a general rule, but it fits what I've noticed in the market. But partners won't work with you if they doubt your ability to win against the competition.

I see partnerships as part customer and part employee relationships. In one sense you need to sell to your partner regarding your product portfolio, support, business growth, and your company as an advantage against their competition. On the other hand, it's also a matter of recruitment, training, monitoring, measuring revenue, customer acquisition, and, in some cases, firing. And if you don't partner with the right people, you won't win deals as often as you should. Sure, you'll make some money, but you'll never achieve your goals if you don't have partners with you. If they're not with you, they're against you in one way or another.

Markets are complex. Each country and each company has local dynamics. Still, everyone can benefit from partnerships. So, what's the best way to quickly create these beneficial relationships, especially if you're a small company or startup with no global connections or an existing player trying to crank up growth or fix underperforming revenue? I had multiple opportunities to plan, build from scratch, and execute on partnerships. I have also had a chance to replace an entire existing partner network, not as a head of partnerships but when I was the head of sales. It requires winning some initial deals (or helping new partners win some deals) to demonstrate your company's credibility and using that to attract partners to do more work. The channel fit or alignment is critical. It's not uncommon to get two thirds of your business from one type of partnership. Nail it before you scale it. These could be consulting firms, global SIs, resellers, distributors, service providers, cloud platforms providers, whoever. Don't spread yourself too thin.

Above all, to land a partnership, you need to be and be seen as winning in the market. That means you're in front of customers closing deals and creating new ones. Potential partners want to work with winners. Second, partners need to know they can win by partnering with you. Then they'll sell your product even if they make less profit margins than they would otherwise. Depending on the nature of your product or solution, partners can create several revenue streams. It takes time to get to the higher revenue stages like training, certifications, projects, and retention. At the end of the day, partners want to grow their own business. Your product brings them one step closer to keeping their customers satisfied and spending. You have to train partners with the same level of intensity and frequency that you train your internal employees. You need to monitor their progress on GTM with the same level of intensity as your own sales teams. Support them as much as you do your own sales teams. If this is not business as usual, you will find partners hard to manage and control because you will be seen as opportunistic. Then you start thinking, "It's difficult running a partnership, so why bother?" and the downward spiral of a bad decision begins.

At different companies I worked for, we've always encouraged our partners to bring their customers to the table. Our pitch to them is simple: sell your customer on speaking to us, then we'll sell them on our product and route the deal through you, a standard approach that works when you are early into the partnership cycle. To instill confidence that this approach works, we sign a teaming and/or nondisclosure agreement. In other words, if you're sharing your customer base with us, you want to know we won't "compete" against you. Once the paperwork is complete, we get the sale going. Together, we and our partner help their customer understand our product, deliver proof of concept, and offer thorough product demonstrations for the sole purpose of proving we understand their needs.

In many cases, if a partner is not working with you in an account, it is safe to assume that they are against you.

Building strong partnerships like this is a specific skill. Most people can motivate themselves. If they want to wake up early and get a head start on prospecting, they can. What about motivating somebody else in another business to work with you? How do you show them you're working toward a win-win? That helping you is good for their personal and professional growth? As with the first two elements of the magical selling system, I will teach you how by example.

YOUR ENEMY IS MY ENEMY

The largest bank in a country needed to upgrade their core banking platform. This customer had zero technology footprint from us. My team and I visited them multiple times, but we had no deal almost a year later. How could we get closer to yes when we were stuck at maybe? The bank was working with a local SI company who was the sole representative of our competition, so we didn't engage with them.

More than a year after repeated attempts to get the customer to consider us, and not making any meaningful headway, I happened to meet this local partner's CEO. It was not a planned meeting, from my end, at least. In that one year, we had won a couple of other banks on our platform, and he must have been aware of those wins. We struck up a conversation. We decided to explore working together for a win-win in the market. Keep in mind that we were *not* direct competitors. During that discussion, I learned that the partner was not entirely happy with the vendor they solely represented. Upon further prodding, we realized that the other vendor was trying to make the largest bank their direct customer instead of having the partner continue to manage the account as they had the past several years. The CEO's reading was that the vendor was after the services revenue, which the partner was bagging along with margins on product sales. I couldn't think of a more perfect opening. It's a classic case of a competitor asleep at the wheel. I strongly recommend every sales rep and manager to be alert to identify these signals, either for a partner or customer.

"Your enemy and my enemy are the same," I said to the partner's CEO. He nodded. We agreed on the spot to work together. Once the paperwork was completed, our new partner leveraged their existing relationship with the bank to recommend our platform as a replacement of our competitor's platform. Prior to this partnership, the bank saw us as outsiders. Like magic, everything fell in line, and we closed a $4 million platform deal. Add-on services totaled $5 million. It was a full replacement of our competitor's revenue. As you can tell, the market for our product category itself didn't grow. It's just that we bagged someone else's big account. Winning back competition means they lose a certain amount of money while we gain an equal amount of money. That's a 2x difference between our competition and us with a single deal. Imagine the momentum this deal generated for other prospects, partners, and job opportunities on our platform. We ensured that there was a generous amount of money allocated to training dozens of people to ensure our products ran properly. Needless to say, we ended up being the number-one technology vendor in that country with revenue bigger than the combined numbers two and three, even though both were bigger than our company globally in several countries.

That was the largest deal for us in APAC that quarter. Our new partner eventually led to us winning six more banks as customers. Soon every major company in the region that used our competitors' products was looking at us. We eventually became the market leader, and this all happened in an Asian country where we initially had a negligible footprint but ended up generating an eight-figure number. That partnership unlocked everything because our strengths complimented one another.

So, look for companies and people whose strengths complement yours. Does that sound easier said than done? Let me give you another example. Years ago, we were trying to sell our software to a government agency. Through our first few discussions, I learned they needed to prevent individual citizens from gaming the system and committing fraud. That translated into identity management software and content management software as the key components needed. The customer didn't have the skills in house to build it themselves, so we found a partner to provide that exact solution built on our platform. How did we win this partner's buy-in? The same way we get a customer's buy-in.

On another occasion, Before partnering with us, an independent software provider who specializes in identity solutions wanted to know if we could lower *our* price to increase *their* margin when we sold our joint solution to a government agency. For a platform vendor like us, giving one company a special discount while insisting everyone else pays the full price goes against the law. The compliance issues would ruin the deal, so there was no way we could agree. Our potential partner wasn't happy. It got awkward. Still, our discussions continued. We learned that they understood the customer—the agency—and knew they needed an auditable, secure software infrastructure to deploy their solution. We provided those functionalities. The partner had their own differentiated solution for identity software, so we pitched our differentiated features. If we could educate the customer on those features' benefits in relation to their KPIs, they would definitely request those features in their RFP. If so, we could get end-to-end differentiation on our platform, plus the front-end system that the partner sold. If we stayed price-neutral on an identical bill of material, the partner could simply differentiate their own product (technically and commercially) to win the deal. We simply used our MAGIC lens. This allowed us to smooth out any friction and educate them on our platform's effectiveness. Our competition had a hold in that region, and other government departments were already using their software. Both we and the partner realized we had to join together to have a winning chance against our competitors. A year and a half later, we won that deal.

Winning partnerships improve your probability of winning client deals too. So, how do you build them?

HOW TO BUILD PARTNERSHIPS WITH ANYONE ANYWHERE

Most sales books will tell you to sell yourself first, whether you're pitching to customers or partners. True, but how do you sell yourself? By establishing your utility to the person who's planning to work with you. But you can only prove yourself useful when you know what the partner wants. Remember the five questions to understand the heart of the opportunity with any customer? Now you should apply them in the context of partnerships. The keyword is CONTEXT. The C in MAGIC. It's a superpower.

1. How do they measure success? Incremental revenue, more wins, customer retention? Whatever it is, find it.
2. What is their measure of that success right now?
3. Where do they want to end up?
4. If they make these changes, how much revenue would they save/make?
5. What is the benefit over time?

Use these five questions to learn what both internal and external partners are looking for. When you look to recruit a new partner or strengthen and expand an existing partnership, look for willingness. You may even do a magic quadrant for the partners you would like to work with for a deal specific engagement or wider engagement with you. Look for complementary skills. Partnerships need to constantly adapt to the deal circumstances. A common mistake I have seen is failing to ask partners these five questions and allowing a misalignment of expectations both ways. Wrestling in inefficiencies. If your company is targeting a certain revenue from a territory in indirect (i.e., deals via partners) business, it is only natural that your partners are aligned with that number and invest resources and efforts to meet those revenue targets. If those two aims are far apart, your ability to get to that quota is left to chance.

There are many possibilities. A partnership ecosystem only works if it serves both parties. The questions tell you how to help your potential partner win. Then you can structure deals so you both achieve your goals through your work together. And you, of course, expect the same of them. If a small-time partner is looking to make $200,000, that means they have to deliver $2 million worth of business (10 percent resell margin). If they're selling service training plus maintenance or support plus products, then probably lesser top-line revenue will suffice.

How do you help this partner find that money? It may be as simple as knowing which of your (or partners') customers are willing to spend that amount on the product or service you sell. These are joint account plans or go-to-market plans. You scratch their back, and the partner will scratch yours. Notice there are different partnership types. The above example is geared toward a boutique partner. If you are dealing with global consulting firms such as Deloitte, McKinsey, or PwC, you can expect them to drive the deals, so your job will be more of a support role. Let's explore other types of partnerships now.

GO-TO-MARKET PARTNERSHIPS

I took up the role of a regional VP of sales at a company in the early stages of setting up its business in APAC. They had a handful of teams at that time. On my first day, the first two people I met were the APAC partner manager and the partner solutions consultant. They were responsible for the partner ecosystem. After introductions, I asked them what our partner strategy was for the region. Their response was word salad.

"It's fine if you don't have a partner strategy," I said. "I'm happy to help craft it as a part of the sales strategy in my region."

This was not the first time I encountered something like that. Most people cannot articulate a partner strategy in a crisp, to-the-point manner. The reason is most people don't think through the process and therefore are transactional, not relational over a medium-term horizon. It is also common to see partner management as a support organization rather than integral to go-to-market (GTM) philosophy. Successful companies in APAC and around the world spend significantly more time and resources to align with partners, constantly and consistently. I'm a big fan of this approach. This can be done in many ways, regardless of whether you are an individual contributor, team manager, or head of a country or region. In fact, partnerships should be inherent to the GTM plan and not an afterthought. If such a plan doesn't exist or is vague, the head of sales is not maximizing the company's revenue potential. That is not sustainable.

We are in the age of rapid and accelerating innovation in every sector of the economy. This is disrupting many existing technologies, vendors, and customers' marketplaces too. Speed and ease of deployment, compatibility with customer's systems, compatibility of skill sets are critical aspects in partnerships if you want to gain competitive advantage.

Several factors determine your channel fit. A big firm typically needs to course correct or expand its partnerships whereas a startup needs to find a right fit. It's also a function of whether or not you are entering into an existing, well-understood market, creating a new cutting edge solution, or simply expanding territory. In the startup world, it's common to have over 50 percent of business come from one type of partner. Establish and strengthen that before expanding to other channels.

If you are a big vendor, you will have to defend your position and grow. You need to do course corrections to your Go-to-Market approach. Your own training is critical, and equally critical is how you bring your partnerships up to speed. Remember, if your partner decides to position your competition's new product or a disruptor's new product, it's a double whammy. It's hard to recover from such blows. Year after year, we had this plan of what competitive strongholds we could convert into wins. Every year we had one or two such "lighthouse" wins. When that happens, it creates a chain reaction. To start with, it helps you overachieve and gets you recognition that will boost your career prospects, whenever you win such deals. Second of all, like a magnet it attracts stronger partners to work with you. They share intel that strengthens your position. There was no exception to this. We had amazing intel into our customers' business, amazing access to the decision makers, their problems, KPIs, and so on as a result of such partnerships. The momentum strengthens your fort and moat.

If you are a new entrant or a disruptor, did you ever check how many of the VC-funded US startups expanded their businesses successfully in APAC? You'd be surprised by how many of them fail. I have observed many of them; each one is very promising and has a lot of customers in the United States or Europe. One common thread you might find is they never treated the partner network with as much importance as hiring their own sales team members. They tend to behave like big companies with a lot of existing pull in the ecosystem, which is totally absent. The C in MAGIC—Context—is missing. How do I know this? Just like most people with a LinkedIn profile, I receive constant messages from recruiters and headhunters. Every time they have a "superman" description for the sales leader they want to hire. One way to tell which company not to join is when they just talk about direct selling to accounts only. The partner ecosystem used to be an afterthought. Sure, there are circumstances under which you would want that role, but don't be surprised if the company closes shop within a short time. The smart ones invest in building partnerships. They hire people that know how to do direct business effectively and also know how to leverage partners in a roughly fifty-fifty mode. I don't mean partner management from an operational standpoint. I'm talking about coselling skills that require constant monitoring to stay effective.

Here are some rules of thumb for you to build such strength and resilience in your business:

1. Homework: Recruiting a partner is the same as recruiting a team member for sales, presales, or any customer-facing role. Allow me to explain. What do you look for in a salesperson? I used to assess their relevant experience, their track record, energy level, network, and solution knowledge. Directionally these are the same things I look for in a partner. What are their stronghold accounts? What solutions do they sell? Are they in a similar space or one or two degrees apart from your solution? How will you train them? What are their expectations on top lines and bottom lines? Sure, it appears complex and it involves a few more steps than simply hiring a rep, but the trade-off is equally good or even better because you get a group of people working to extend your ecosystem.
2. Readiness: Once you hire them, how will you bring them up to speed? A lot of companies have good internal training modules but comparatively much fewer for partners. You need to increase that to a proportionate level. I'm not suggesting that partner reps will spend exactly the same amount of time learning your product as your reps do, but they still need active engagement with constant intensity to get top-of-mind recall of your products and services.
3. Ongoing management: An ideal setup would be for the company reps to have a direct line of connection with the corresponding account manager in the partner organization to work on jointly. One of my favorite ways to do this when I was in an individual contributor role was to call up the partner reps dealing with accounts or opportunities in my territory starting between eight thirty and nine o'clock to find out what they were planning for the day and then follow up with them at five

thirty. That way you have a chance to help them where needed and work as a team. It is possible that you don't have so many partners or so much business in your current situation. If this is so, do the same time that you would have to do with the sales team. Whatever the objective, ensure you or your team speak to your partners on it as often as you speak to your sales reps. The big benefit is a virtuous cycle. Once we began making strategic partnerships, my involvement with a customer typically lasted until I ensured we had a technical win against our competition. Partners would help through this part, but I was fully involved. From the time I realized it was our deal I would leave the rest of the process to the partner. They were good at paperwork anyway, and I didn't interfere. While every deal can be different, generally speaking, in my experience the partners who are stronger end up getting more margins. That frees me up to chase other opportunities. My primary goal was to drive away my competition as much as possible. Saving time was always high on my agenda. If we quibble on 1 or 2% of the margins, we lose valuable time taking eyes off that goal. Partners also tend to be very understanding on the margin front and flexible as long as they are retaining revenue generation potential like services or can cross- or upsell to that account. Giving away the paperwork part or deal closure part would free up my time to get more deals to the table. This provides leverage in your negotiations with partners either on price or in a give-and-take in return for another lead that you have been driving while they were closing your deal. It's a virtuous cycle.

4. Measurement: Not all partnerships or situations are conducive to a quota as in the case of a sales rep, so find a mix of new logo acquisition to services to hard revenue targets that works well for both parties. It's a combination that helps partners win their business against their competition; retain their existing customers; or upsell or cross-sell to make it a functional, win-win relationship.

You can apply the same magic lens to partnerships just like you do with customers and build a skill-will matrix for partners just like you would build one for your employees (see chapter 5). Something that I didn't notice most startups do quite well. Some companies I noticed over the years start with partner management only and not with sales. When you say partners you need both parts to work well. Your own sales team and the partner organizations. Missing any key ingredient and you'll wonder why the dish doesn't taste so good anymore.

Partnerships come in many forms. There are transactional partners, SI partners, technology partners, service-based partners, etc. Which one fits your business in a certain territory or account is for you to prioritize, but irrespective of the combination best suited for your business, I recommend you look for the following (as shown on the y-axis in the picture below):

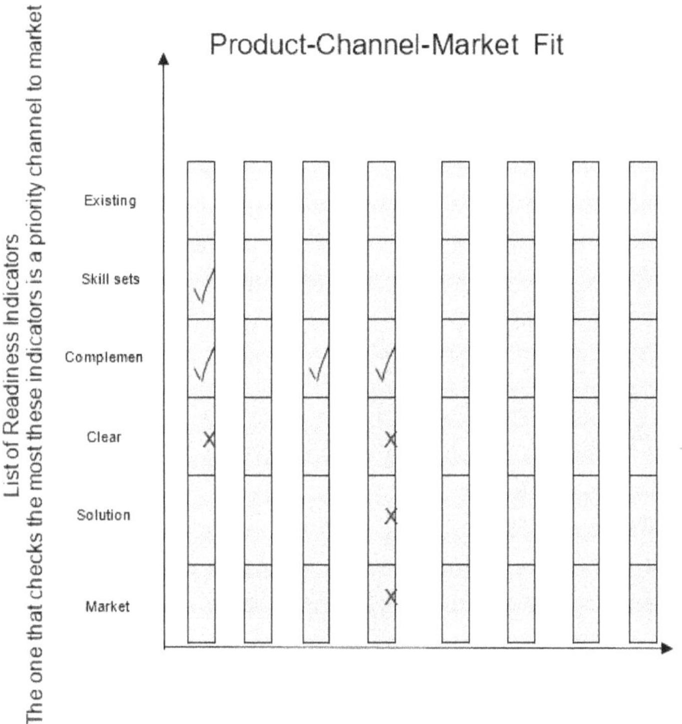

1. A winner with established stronghold accounts that match your own target market. In a way this is equivalent to an in-

terview with a potential candidate or a deal qualification for a customer. I'm aware of a case where the startup got $2 million in an unlimited agreement but the partner ended up getting $4 million in margins on a total deal value of $6 million. The partner got more revenue than the vendor. Clearly the partner had the skill and the knowledge of where to sell.

2. Solution alignment. They have existing practice selling/servicing in the same space and aware of the sales cycles. Or they are one degree away from the solution area that's easy to learn and expand into.
3. The key aspect is to set expectations and understand expectations in terms of metrics. Just like in the case of customers, the majority of reps and partner managers don't discuss hard, quantifiable numbers early enough. Getting to this takes time. The point I'm making is that both parties need to know what the other party is expecting out of this partnership and if there is a plan to get to that. The easiest place to start is with one or two accounts.
4. Complementary strengths: In some countries, you need to have local language partnerships for level one support. It will also involve converting all literature, websites, and so on to be tuned to those markets. Partnerships are one of the best ways to start this in any country.
5. Building partners' skills and regular pipeline generation and deal progression activities need to work in tandem for a partnership to thrive.
6. Have a matrix as shown below. There is a way to calculate the scores to plot this graph and keep it updated as and when changes happen to people movement, competition presence, customer dynamics, and so on.

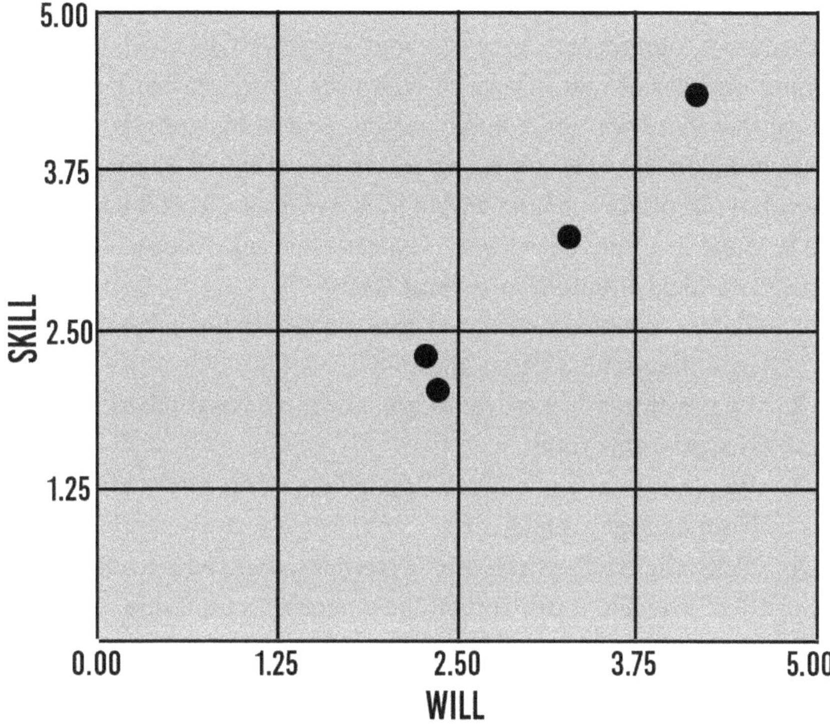

Figure 7 Skill-Will Matrix.

Sometimes you have to make a parachute on the way down. Not all the time do you have the luxury of a fully set partner network in your territory or account. Many times you have to create and manage a relationship with a partner during the sales cycle.

A couple of years ago we received information of a deal being constructed in a global insurance company's Asia office. Our company had been in the region for less than a year versus our competition who had set up shop some five years before us. They had spent several months in that account by then and completed a successful POC. Our account Manager reached out to the insurance company's person in charge of this evaluation. We were told that they had progressed too far and couldn't give us an opportunity to participate. It was clear that he was like our competition's champion. Our attempts for the next three weeks didn't yield any results, so we decided to identify which partner would be interested in working with us in that account. After trying five different partners, our account manager met a partner CEO who knew the managing director of that insurance company. Through that partner we got word that there was an evaluation going on in his company and legitimate competition was not being invited or allowed to engage. The MD assured our partner that he would look into that and that the process would be done in a transparent way. Experienced salespeople will know that this is an ideal situation in several ways:

1. The competition educated this customer for several months including with a demo and POC.
2. The customer's key requirements must have taken a proper shape by this time.
3. We didn't have negative baggage and therefore were starting from neutral ground.
4. Since the competition didn't have anything new to show after POC, we had a free run on the customer's undivided attention for the next few weeks/months.

Conversely, if you follow the magical selling process you would avoid your last-minute competition getting ahead of you after you already spent tons of time and energy nurturing the customer.

We proved that our product was a better fit. Based on the evaluation of the technology the customer decided for a direct negotiation. The customers liked us so much—thanks to the magical selling process executed well—that they decided to call for a Named RFP. This means that they asked for a specific vendor product, and anyone proposing needed to bid only for our product. From being an outsider who couldn't find a way in, despite all our sales skills and intentions, we turned it into a named tender and a quick win thereafter. The situation went from a very low probability of winning—because this was a net new logo, with no relations, no partners, late entry into the market—into a guaranteed win. The linchpin in making this happen was the partner we recruited in the middle of a sales cycle.

The bottom line is that partner selection for a market or an opportunity as well as going-to-market planning are critical. We don't always get it right, and that can easily mean the difference between winning and losing.

Sometimes you are lucky enough to get good cards. You have a product that sells by itself, and all bidders—no matter how strong or weak they are—don't have a choice. In a competitive world, though, you need a network of partners.

SIGNS OF A WEAK PARTNER ECOSYSTEM

The following red flags will save you so much grief. You can take my word for it.

1. The weakest partner ecosystem for any vendor - other than working with your competition - is when partners simply respond to a transactional request. They are riding on the pull your product has generated in the market. This is a one-way street. Fix it before it's too late.
2. A partner ecosystem when Inspection (I in MAGIC) is weak. Inspection of activities, pipelines, account plans need to follow a success cadence format. Don't take it for granted.
3. If you find yourself talking to your own team and some direct customers for much of your time regularly..
4. In rare cases where you are managing an account and your solution is dominant, you might feel you have an advantage you can ride on. Missing to strengthen partners in good times. You will have an unfair advantage when you have several information sources on ground. Have your own customer success team team on the ground to ensure the technology is used properly to the maximum extent. Outside of this it needs to come from partners.

A change in your company strategy or a disruption happening in your field needs to be treated as not just for your company but for your company's ecosystem. The course corrections also need to be with that view. I have seen quite a few situations in my career where a disruption to one's business or even simply weak results in sales are a result of not recognizing the problem as an ecosystem problem. Wrong diagnosis leads to wrong medicines.

On four separate occasions in my career, I was assigned the responsibility to fix underperforming territories. If I generalize it, revenue drop in each scenario was around 10 percent year over year for two or three years even when the market for those solutions in general was growing in the mid to high single digits every year. On one occasion, when I asked my boss, what he wanted me to do in that territory, he said, "What you usually do." That was a two-second brief. He didn't tell me what was wrong with the territory, how he wanted it fixed, or anything like that. Each of those four situations were very similar in terms of symptoms: the customers were not really engaged, they were spending but not on our product, customer-facing time was with wrong personas. This meant no time with economic buyers. If I want to meet several customers in the new territory the country sales director takes me to lunch with a procurement manager, you don't need to be smart to realize where the problem is. Another sign is when the partner manager or channel manager doesn't have a clue what the country sales strategy is. It's a gigantic waste of everybody's time. This is a clear misalignment of direction, activities, and therefore outcomes. We were using a hammer when we needed a screwdriver!

What did I focus on in each of these cases? For one thing, I looked into a short list of customers who were spending on other vendors to have direct engagement with; the agenda here was a quick win to show that we were a winning team. You will find these situations in any running business. In parallel, I wanted to work on understanding the partner landscape: who's active in which accounts, who is winning, who are they partnering with, who is having a bit of a tough time, etc. You can easily get this intel in two to three months in any country while you size up the customers' needs and learn how your solution fits those needs, in parallel. At the end of those three months, we had to bring in partners for specific opportunities. The green shoots usually started showing in the first quarter in terms of fresh energy levels; in the second quarter in terms of active opportunities; and in wins in the third quarter. In each of those roles, by the third quarter out, we had solid partners who were prepared to invest in working with us and in growing our business.

One specific example: The fact that we started with a large customer cancelling a half-million-dollar contract and an active plan to get out of our company technology just when I arrived, to no new purchases for the next ten months, to giving us $14 million in the following eighteen months, didn't happen all by ourselves. We began to turn around those after ten months with direct engagement. Partners active in that account, but working with our competitors, started noticing that we were winning and gaining customer trust. They reached out to us to work together. My question in validating their capabilities was, as previously stated: "If we are winning on our own on this account, what additional revenue can we make by working with you?"

The partner's response was crisp. "You guys are good at what you are doing, but you don't have all the intel on what other projects are active in that account."

Partner shared specifics of 2 deals. My account manager and the sales director were not aware of those. I was certainly not aware either.

That discussion didn't last more than twenty minutes, if memory serves, and we got a solid partner out of it. I would say about 40 percent of that $14m revenue wouldn't have come to us if we didn't have a partner.

SIGNS OF A STRONG PARTNER ECOSYSTEM

Just as there are potential partnerships warning signs that tell you who to avoid, there are also green light signals. Here they are.

1. Constant, verifiable intelligence on a customer and/or market landscape
2. Interaction every day or two over prolonged periods
3. Engagement in creating an opportunity instead of just tenders/ RFP responses
4. Trained people offering services as a revenue model for that partner
5. Constant pipeline generation activity

RECOMMENDATIONS FOR RUNNING HEALTHY PARTNERSHIPS

I always see partners as part customers and part employees. You have to sell to them why it's win-win working with you and also manage them as they end up selling to your end customers. Any plan that doesn't include this will make your relationships vulnerable to competition.

Partnerships with complementary solutions: The main thing we look for in any partnership is to have complementary solutions or strengths. This could mean service skills, account relationships, or technology or a product that complements your own technology footprint.

Creating deals together: Start from PG activities and move forward through deal progression. Nothing less than 2-3 touch points every week.

Constant upgrading of skills: That means product updates and learning what they mean for the customer. How many webinars are you hosting for your partners sales and presales teams?

Alignment in outcomes: We are not looking at perfection in every partnership. Context is the key for either one off or long term partnerships.

Course corrections: How often do you do course corrections on your deals or with your colleagues and team members? Are you getting optimal results? You may be an individual contributor or a sales leader. Some prefer doing quarterly business reviews and regular cadence meetings. Personally I prefer to have ongoing conversations with partners for enablement, pipeline generation, deal progression and forecasting at least once a week with my counterpart.

There are any number of forms of partnership, and these are further evolving. Let's take the cloud vendor, Amazon Web Services (AWS). It is a leader in Infrastructure-as-a-Service (IaaS), and it offers a variety of software platforms. Microsoft Azure came in from behind and is the market leader. Overtaking AWS not an easy feat for anyone. What do you think is the reason for their success at that scale? There are many reasons you will get depending on who you speak to. About 2 years before Microsoft Azure grew past AWS, I got to know from more than one partner in APAC that working with Microsoft is frictionless. From single point of contact for specific, to simple ways to access partner development funds to clear referral arrangements (or incentives) were fully aligned with each other. Azure scored a lot more in all my anecdotal discussions. Anyone will argue their product and technology is better, but for such large, mature corporations many times it's the partners' weightage that swings the market share. No wonder Azure is pulling ahead.

Not every company is at that scale. What happens if you are a small company? Many small companies or startups want to grow fast and focus on—of late—smaller deal size, thanks to cloud subscription models and quick sales. I have spoken to many CEOs of such companies. They tend to use the big-company approach of trying to get a premium and end up spending enormous amounts of time on sales cycles and negotiations. In theory the approach looks good, but practically the trade-off on time is not worth it. Get some wins and focus on partnerships.

That's why you should start with a plan to enable and certify resources, not just the global SIs or consulting firms, which can take a bit of time to work with, but also boutique SI firms with specific strengths and insights into a territory. If they see that traction move, it won't be long before your pipeline starts to do so as well. You have to nominate the partners, run with them, plan with them, and win with them at least half the time. Follow the methodology and maintain a constant engagement with live tools, documents, and reviews.

As I mentioned in the example between Azure and AWS, how you organize within your company in terms of KPIs has a direct bearing on the effectiveness of partnerships. So for partnerships to function well externally, the management of partnerships internally needs to reflect that approach.

Ensure you bring flexibility to partnering too. It is a good idea to occasionally do direct deals also—that is, without a partner. I deliberately keep it on low but on all the time. Why? Because direct deals keep me close to the ground. They get my hands dirty and keep refreshing and fine-tuning my skills. When we do direct deals we end up maximizing the deal revenue, but more importantly allows you to land lead partners. You need to plan for those trade-offs. Many partners tend to respond to people who know how to win in the market, and if you are one such person, then they allow you to coach them too. They know your suggestions are practical and you can jump in where required to help them stay the course. You have the required skill to either lead or support the partner depending on what the situation demands. A leader is someone who goes the way, knows the way, and shows the way. This applies to sales situations too. If you don't go the way, you may not always know the way, and it's hard to show the way to others. So keep practicing deal closures and every part of the sales cycle. Do it in a manner that pushes the existing boundaries of every deal. Try bigger deals, better combinations, different products, whichever angle you can find. Pushing boundaries requires skill and maturity to handle successfully.

INTERNAL SELLING 101: BUILDING PARTNERSHIPS WITHIN YOUR COMPANY

How you are engineered internally within your company has a direct impact on how you can pull off great outcomes with partners externally. When it comes to enterprise sales, partnerships within a company are usually made up of five to ten stakeholders. There are colleagues from the channels team, the presales team, the architects, the industry solutions specialists, the salespeople, and sales management. Everybody is looking to win deals. The way to make internal partnerships worthwhile for everyone is to align strengths and KPIs.

If you are new, young, or responsible for a smaller line of business, you might be worried that internal partnering will be particularly difficult. All you need to do is build a conversation in your own language to answer the five questions at the heart of the opportunity. Ask for help from team members to get such information when they meet the partner or customers. You'll see that everyone tends to be more productive when you do. Why? Because of alignment. When two partners are aligned, the relationship is frictionless. Are you aware of any deals where different team members think of different approaches to win the same deal? I've seen many.

So, how did I get the internal help I needed to move my deals forward? Some might wrongly think playing politics is the only way. That probably helps in companies where employees went to the same school, belong to the same golf club, or go out together on the weekends. However, the tips I provide don't require social status, privileged connections, or movie star charisma. You could be an incoming, experienced joinee or a junior salesperson these tips would get you the internal partnerships you need.

Years ago we were targeting a government deal in Southeast Asia. I was the GM of sales for the region, and we did work up to the point where we needed somebody with subject matter expertise to move the opportunity forward with a particular set of people in the governmental department. It was a large deal (approximately $3.5 million), and I started looking for a senior person. We reached a VP who had joined our company a few months prior. It was my first time talking to him over the phone. He would listen but gave me the impression that he was busy travelling to Germany, Japan, and other "important" places. They were not on my list of important countries as my territory was different. He would give me a time and then move our meeting when we got close to that date. This happened two or three times, and I was pissed off that he was not keen to travel to the emerging countries. (I'm sure he had his reasons, even if they were worthless to me). What we really needed was someone "senior" who could speak the customer's language. I reached out to a VP for solutions and knew quite a good deal about this subject, though he was not a government specialist. He carried a VP title and that could prepare me well to present to the customer. That ended up working extremely well, and we won that deal and went on to make several other memorable deals.

The point I'm trying to get to is that the government SME assigned to my region was not cooperative. Like I said, it was a $3.5 million deal and was one of the most high-profile government deals in that year in APAC. A month later I got to know that the same guy who refused to support us ended up presenting this deal as his own deal and got a lot of accolades, taking credit for what he didn't do at all. It was pathetic behavior.

The following day my boss and I were on our way to the airport to return to Singapore, and I relayed the story of how that guy was taking credit though he had in fact refused to help with the sales cycle. My boss asked me to ignore his behavior and let everyone take credit. In fact, I should give more credit to others, if anything, he said. It was advice that I ended up taking seriously and practicing later. These situations where someone who didn't work takes credit are common, and I'm sure you have experienced it in your career many times. I practiced it with a twist, though. I keep a tab on everyone who is contributing to any opportunity or territory. When some work hard and some other person gets or takes the credit, the one who is really working is the one who gets demotivated to work hard again or may even decide to find another job if that continues. That wasn't good for business. Thanks to my boss's suggestion, eventually I would tell all the sales reps in my team to plan to give away credit (and keep their commission!) as an advanced planning mechanism. This is more true for a sales support team. This helped in a way of setting expectations so the reps didn't feel bad but were vigilant enough so they see it as a "leader" behavior. When they are willingly planning to share or give away the credit, it's hard to feel like a victim at the same time. Why is this important? Team members respond better and will align with you better when they have these expectations up front. The end result we are looking for here is to have successful selling and maximum wins, and this behavior helps build and/or strengthen your internal partnerships.

QUALIFY THE DEAL FIRST, BRING IN A PARTNER SECOND

Years ago we were up against a deal that had two separate market leaders, one in data warehousing and the other one in data mining. We were in an account where we didn't have much of a relationship (we had come in new) even though the customer was a big user of our software. The customers eventually decided to do two separate projects instead of the original one. They started with data mining first. Our position was much weaker in mining than in data warehousing. We decided to go in and expend effort anyway because a) it was a small deal and winning or losing was not going to make a material difference to my numbers and, more importantly, b) I wanted to get to know the customer's decision criteria and decision process. This would help us better plan for the data warehouse deal, whose value was ten times bigger than the mining deal. I told the architect involved, that I couldn't see a way to win the mining deal, but I would consider it a win if we understood the customer landscape and he was clear on what needed to be done for the warehousing bed. As estimated we lost the mining deal, but we won big on DW later that year. Engineering is the activity you do, and magic is the result that comes out for everyone to see.

One of the biggest issues in running a business I have noticed is when sales leaders are not close to customers and instead focus on operational management. If that persists in your management, either find a way to correct it or simply find another role. These kinds of leaders are usually very good at managing upward and taking care of themselves, but they are neither competent nor willing to learn. I was once being interviewed for a job and made it to the final round, but the decision went in favor of someone else. I was shocked to learn who they hired. I knew that person, and they had had compliance-related issues and were known to generate friction in the market. How smart people couldn't pick up on those issues during the process is baffling. The people running those companies were amateurs. To be honest, I didn't pick up that weakness in the hiring manager and I felt lucky that I didn't get that role.

APPLY "SUBTLE" PRESSURE

Winning over employees and management to help you is often easy, but there will still be instances when another person or project is prioritized over your deals. This happened to me with a solution sales head. He would just ignore my requests, so I decided to apply some subtle pressure, just enough to get his attention. Too much pressure, and he would push back. Too little pressure, and he wouldn't move. The art of subtle pressure is an important skill to develop.

After asking him for help on my deals three times, I went to him in person. "Let me get this clear," I said. "After three times of asking for your help, you're not supporting me. On my QBR (Quarterly Business Review), I'll have to put you down as the reason for me not progressing deals in the pipeline. I'm giving you a heads up for you to be ready to respond"

This let him know I was serious yet not disrespectful. He would be held accountable for those lapses. I didn't want to throw him under the bus, but I wanted him to do his part. After that interaction, he moved the resources I needed to close those opportunities.

Will subtle pressure always work? In fact only subtle pressure is sustainable for any meaningful length of time. Whether it is partners, customers, managers, subordinates or your colleagues, subtle pull or push is what gets things moving. You may have to dial up or down what is subtle for self motivated individuals vs low energy individuals, but i found a really useful tool.

What's on the Truck—The Eye Opener

I should also warn you that if despite you trying to toe this line you don't succeed, you need to be sensible enough on what amount of time you are willing to put into this. My approach was this: 70 percent of my time goes to everything customer- and partner-related and 30 percent goes into building these internal alignments. Anything more than 30 percent needed to be a temporary thing. This was a tipping point for me to quit a company I enjoyed working with. This happened when they hired a new head of APAC business, and he was a pure operational manager. They started by dismantling things that were aligned to customers' needs. Our real strength came from a) working products and b) aligned relationships among all frontline teams and between us and its customers. They decided to cut the second part completely.

I remember the day I decided it was time to move on. This happened when the new, "no good" guy took charge, and in my usual enthusiasm I decided to present the way to generate $100 million in stand-alone cloud revenue in Southeast Asia. I gave four examples that were in my pipeline, one of which we won and all of which had nearly $1 million in total contract value. I made a few asks in that meeting—some twenty members of senior management were present, most of them didn't have APAC experience—on what was required to play to our strengths. I was almost expecting that it would be an eye opener that there was a path to win that revenue. Instead the response was, "Look, all this is good, but you need to sell what's on the truck."

What's on the truck? I thought to myself. *You can hire a bunch of people to do that. What do you need me or my team for?*

That was the first time I had ever heard someone make such irresponsible statements when it came to the early adoption of any product. It was also my worst presentation. On that day I realized that, with that management in place, it was a matter of time before it became a complete train wreck. A little over a year later, I moved on. I was not willing to set aside any more than 30 percent to do internal management. Anything more than 30 percent meant that I lowered my focus on my customers and their needs. When you are in sales—and I don't mean you are a rep or a team manager—even if you are managing hundreds of team members, you must stay close to the ground. When management doesn't pay attention to feedback from the ground, you know it's time to move. That move turned out well because it provided me with the opportunity to look at things that were moving quickly. It also exposed me to several other ideas, like setting up new teams and businesses in APAC. I started my own business and also wrote this book to help customers connect with new technologies quicker and help promising startups grow their valuations quicker. This is even more enriching than I expected, but I won't forget any of the lessons I learned at a dozen different roles I played.

MANAGE YOUR SALES MANAGER

Did you ever think of treating your manager as your customer? You need help internally to get your deals through. Sometimes it's pricing. Sometimes it's changing the license structure. Sometimes it's adjusting the GTM plan. Other times, the help you need is not directly related to closing a deal. You might need to get a budget approved. Maybe you want to negotiate a pay increase. Whatever the situation, in my experience, internal selling is an afterthought for most salespeople. But it's no less important than meeting the customer.

Why? You may put yourself in a winning position with that customer, but what if senior management doesn't want to approve the deal? The obstacle could be your boss, an operations manager, or someone at the head quarters. There are many reasons why you might feel your company is not supporting you. Sometimes the circumstances are no one's fault; other times incompetence is to blame. In either case, it is in your hands to fix it.

Before I joined Oracle, I worked for smaller, well-known companies in India. At times, I simply couldn't move deals. If you're a decent salesperson, and you're unable to close, something's wrong.

I was finding it difficult to move the sales cycle forward despite my best efforts, so I took it up with management. At a small software company I worked at as a young salesperson, I decided to ask our president for his feedback on why my deals kept stalling. The president was a highflier. He graduated from the best business school, the ones where you have a one in 100,000 chance of admission. It was during a meeting with all fifty salespeople that I brought up my concerns.

"I don't know how long the tough times will last," the president replied.

That one-liner surprised me. It showed his lack of understanding and unwillingness to address the issue directly.

No true leader ever gives lousy answers like this. That's when I realized this guy didn't know half as much as he claimed. Here was the head of the company, and he had no answer beyond, "Go figure it out yourself. I'm just here to count the beans." This type of operational management only wants you to close deals, but they don't address your needs.

Regularly have your manager meet your customers' executives and ask for help with the deal progression. And keep your manager informed about what's going on in your territory - not just good news, but where you need help with the bottlenecks - in a manner your manager can describe it to their managers. Making it predictable for your manager is the best way to leverage him or her for your success.

SET CUSTOMER EXPECTATIONS

"Hope for the best but plan for the worst."

This familiar advice is useful when your deals depend on internal buy-in. Try to plan for and predict the resources you need, even months in advance. Ask the internal decision makers whose support you need if you can block out their time and/or resources. The earlier you do, the more easily you can defend your request. This depends on the deal you're going after. If you have a big account you're trying to close, you know you're going to need help from your internal teams to fulfill the order. Ask for that help before you need it but after the deal has been qualified. This also buys you time in case there are delays in the deal process. Whatever timeline you expect to stick to, extend it. If you think it will take a week to get the information you need from your team, tell the customer 10 days. If you do get it in a week, it's a pleasant surprise. Compare this to an unexpected setback. Nobody wants to deal with people who miss their deadlines.

I'm very particular about setting expectations I can beat so that both customers and internal partners see it as an over achievement. This only works when you plan in advance. Always aim to have cushion, and you'll get to wow the customer with your efficiency, which differentiates your company from the competition and increases the likelihood the customer will buy from you. Another way to negotiate in a frictionless manner is to update your team (or manager) in advance what you intend to propose to the customer as a way forward and check their willingness and readiness to get that work done. This allows you to play on the front foot since you already have the approval in hand. Customers appreciate sales reps who deliver on their commitments promptly.

LET THE BOSS WORRY ABOUT PRICE

Every enterprise salesperson finds themselves in an awkward pricing negotiation sooner or later. The customer wants a deal, yet you're not authorized to offer any discounts. Is there a solution you haven't found yet? Is there someone you can ask to help you? If you're new to your job, it's a new product, or you're not very experienced in sales, bring the customer's request to management so they can help you with the negotiation. Don't promise to drop the price yourself. A manager's gun is bigger, so they might be able to get it done. Discount discussion is typical in enterprise sales negotiations.

Experienced salespeople may use the end of the quarter as a pricing negotiation tactic for both the customer and their company. You and your management both want you to reach your quota. If you're falling short with a few weeks left, giving the customer a discount may close the deal and achieve your quota. You can also use this tactic to impose a deadline on the customer. Their price, your timeline. I use this practically every quarter.

Before you offer a discount in exchange for a faster close, find out if closing the deal that quickly is possible. Can the paperwork be processed on time? Customers might say they can give you the purchase order within a week. Sometimes you doubt whether they'll be able to pull it off, even though they're trying. One missing signature could stall your deadline. Use your own judgment. If a customer agrees to get back to you in one week, do their intentions match their words? Think about who you're speaking with, and what purchasing process and approval chain they have. This goes back to understanding your customer's process.

Only when the customer agrees to the deal do I go to my management and let them know the situation. Far more often than not, my management agreed to the discount that the customer and I discussed. Ask and try justifying the lowest price internally and the highest price externally. For pricing, timelines, activity, etc. This is Negotiation 101. An experienced salesperson never starts negotiating at the number they want.

Discounting is often a winning tactic for the salesperson, but not for other departments. They lose money because the final sales price is lower than the amount they accounted for in fulfilment. In any case, set an expectation with your boss for the worst-case scenario of when and at what price you expect that deal to come through. That means you have the boss on your side to defend the discount.

When fully developed, your internal selling skills will earn you special treatment. Several years ago, I had a VP who always approved my deals. He wouldn't approve the same discounts for other people. Not even for my boss, who was also a VP (same rank). There were times when he would go to the VP to negotiate on my and other salespeople's behalf. The answer was no. If I went to the VP myself, I'd get approval without difficulty.Temper all my advice on discounts with legal compliance.

With the recent developments due to COVID-19 pandemic, pricing is an important point of leverage. So are partners. Based on your situation and the impact to your customer's business, this may mean discounts, but what I'm really suggesting here is "flexibility." There are some established businesses that don't need much price elasticity, but for the vast majority of accounts and vendors, pricing flexibility is a very good opportunity to get closer to Act 1 (understanding the customer) and Act 2 (ensuring the customer knows you understand them better than all your competitors). Companies are trying to keep their lights on, and therefore they are bound to look at how they measure success (heart of the opportunity). Even if they had a measure earlier, check with them to see if it has changed and, if so, what the revised measure is. This change is what you need to leverage to come up with a pricing strategy.

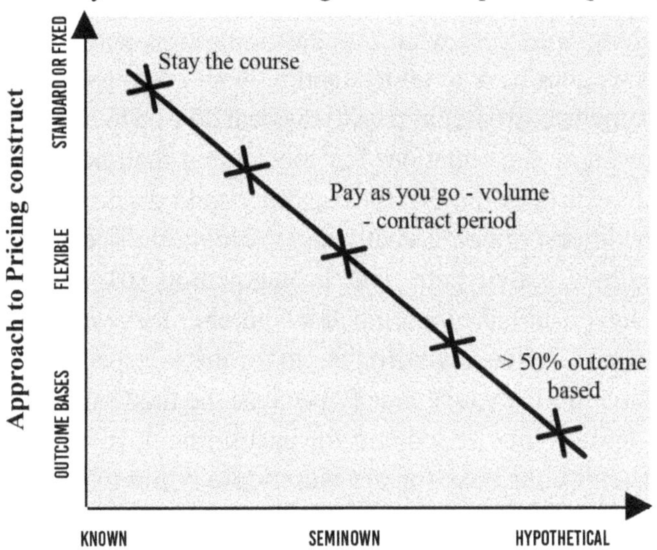

Figure 8 Approach to Pricing.

We did that in the month of April 2020 with our key customers. You not only get to understand the customer by staying close and curious as to how their business is impacted, but when you get back to them with a revised pricing strategy based on that impact, it will show you in the good light of being a real partner. In our case it involved coming up with a new metric of licensing in line with exactly what the customer cared about. We then added to that a pay-as-you-go plan—just on that metric—and the usual volume-based and long-term contract-based pricing. It is the change in metric that clinched the deal because in a way the customer saw this as outcome-based pricing. That's what this graph tries to demonstrate. If there is no option for the customer but to pay you the full amount, good for you. It's your call. But for solution areas that don't have such a clear ROI or measure of success, especially those of innovative and collaborative projects that involve validating hypotheses and several iterations of work, come out with either a mix of fixed price and variables based on outcomes or mostly variables. Customers typically appreciate this approach more than a simple discount on your existing license metric, analytics, and so on. A simple discount may be construed—in some cases—that you may have charged too high earlier. If you are competing against anyone like this, take the flexible approach in order to get a first right to refuse. In any case, if you are a small business, it's better to target cash flow than margins in such uncertain times.

ASK FOR HELP

Everyone has ideas. Internal selling also involves asking for feedback from trusted partners, whether they're the sales director, an account manager, or a sales rep new to the job. Assume that they may have seen something you're not aware of. Asking for help is respectful, objective, credible, and it shows integrity. You're building bridges, not pushing people around with your ego and calling all the shots. Requesting input on your deals is also a way to figure out who else can contribute to your efforts. Explain the current situation and ask what others think. Say you have a couple of ideas, but you want them to weigh in. Have an objective discussion about what's important to the customer and how the competition may be positioning themselves.

It goes without saying that you'll get different, contradictory ideas from different people. To keep colleagues from arguing over each other, ask people individually rather than in a meeting. This way you don't have to moderate everyone or be the alpha in the room who shoots down all but two or three ideas. Still, don't set the expectation that you'll follow whatever idea someone has. Discussing pricing in a group is always difficult. As far as possible, do one to one discussion with each stakeholder. You want their feedback on how to maneuver your deals toward closing, and you're looking at ways to get there. Let them know you appreciate their help and will validate them by speaking with the customer to see which idea works for them.

If someone's idea didn't fly, be sure to give feedback objectively. At the end of the day, it's about winning the deal. Whose idea it was is secondary.

BUILD ALLIES AND RECIPROCATING YOUR SUPPORT

Whenever you are in a position to help anyone, just do it. Don't lose that opportunity. It's an incredible way of energizing yourself and staying grounded in objectivity. When a business development specialist asked for my help, I said yes. My job wasn't at risk, but he was under pressure. Together, we got his proposal finished and got him a job. You can build allies by helping others when they need help. Reciprocity is a big part of internal selling. Give first. That's why they call it give-and-take. Life is about giving, not taking. Just help. There is no need to keep account. Reciprocity strengthens partnerships. Will the people you help always do you a favor in return? No, but in my experience, there's about an 80 percent chance they will.

Never underestimate reciprocity and the benefits it brings to your table, but don't go in with an expectation of what you'll get in return. Keep that reciprocity with as many people as possible to overcome any challenges you may have, but don't make a big fuss about it. You shouldn't have to bring it up because the people you helped will remember. The only thing that matters is whether you are relevant in any situation with customers, partners, colleagues, potential employers, and so on. Reciprocity propels you to relevance more than you think. It's a quality that helps you implement Magical Selling easily in your day to day activities.

MODIFY AS NEEDED

Remember that context is a superpower. Sometimes you have to use a combination of these tactics to get the internal buy-in you need. Every situation is different. In your case, one or two of them may not work. These ideas don't solve everything. There will be countless situations within the same account in the same year that are all different. Think on your feet as you A/B test these tactics. See what works for you and build on that to let your skill set grow. "Nail it before you scale it," I always say. Modify my methods to fit your situation. No sales tip is one-size-fits-all. Internal and external partnerships as a part of your daily engagement allows you to see things a lot earlier than less skilled salespeople. As you start working with this approach, you will soon notice that you are able to see the outcomes so far ahead that it approaches magic!

ACT 4

CHAPTER 5

Win with a Plan

"Customers decide the competitors for us. Not we!"

A win plan is like a GPS that you need to get to a destination—closing the sale—from wherever you start your customer engagement.

As with a map and its roads, streets, highways, traffic conditions, peak hours, weather, speed, on-ramps, and off-ramps, there are many variables in your win plan. Variables such as customer, market, partners, competition, your own team members, product maturity, and other deal-specific situations are essential to consider when writing your win plan. Get the map right, and everyone gets where they want to go without undue detours.

Recall how we divided MAGIC into a Thought Plan and an Execution Plan.

Figure 9 Your Magical Rating.

Let's take a hypothetical situation to see how MAGIC manifests in the execution of the win plan. If you get to know a deal and you are asked to present what are your chances of winning, let's see how the conversation goes.

In this hypothetical example, just give a rating for each letter from 1-5. You met with a client today, and we know who's running the project and decision makers. Be conservative, if you are in doubt. Even if you cover the MAN very well and rate 4 or more, what if a year ago we had some issues with customers, and they replaced our platform with something else in a different department. The context will score less, a 2 or 1 out of 5.

A simplistic calculation will make this deal score 16 out of 25, but the weakest link in the chain is at 2 out of 5, which is below the acceptable level. So we can list actions A, B, and C in order to improve our scores. Anything below a 3 is definitely a red flag and requires a specific plan to strengthen that link. In order to improve the score, we need help from so-and-so for such and such work within a certain date. I want you to look back at all your important wins and any important losses.

Most people don't draw up win plans. They stick to the typical questionnaire or product demos that many times are insufficient and sometimes irrelevant. They are talking about the drill bit. Creating and sharing a win plan reveals a clear path to victory, no matter the country, the opportunity, or the product. The players tend to respond with top-notch preparation. They actively contribute to moving the ball along. They also tend to let you know about any red flags they see that might impact their (or your) work. Without the clarity that comes with a win plan, the team tends to be unproductive. If you're driving in the wrong direction, going faster does nothing to increase productivity. A win plan provides accurate direction, and with clarity comes effortless speed.

On one occasion, we approached an insurance company that was speaking with one of our competitors for several months. The insurance company flat-out told us they weren't interested in evaluating our solution as they had gone too far with POC for several months and we had never shown up for that evaluation process. They didn't have much time, they said. At that stage, most people would be tempted to drop their pursuit and find another prospect. Some may even try to submit a bid anyway at some ridiculous price and/or simply sabotage the deal. When the account manager told me this information, in my mind the win probability for that account was less than 10 percent (based on the MAGIC lens demonstrated above).

We thought it over and agreed that on our own we can't win it. We needed to find a way to get to the table (i.e., first get our win probability to 20 or 30 percent and do things to increase the win probability first to fifty-fifty and then all the way to 100 percent). As you can tell this was a real opportunity but it's important to know what stones to move to make headway. The variables we considered were a) this was a new customer for both vendors; b) the big data platform solution we were selling was just beginning to form the ecosystem; c) competition had a head start to shape the deal for them; and d) our company had just set up operations in Asia about a year prior to this whereas our competitor had a five-year head start in the region. We just needed to find a new way back into the company. How could we recapture their attention?

We decided our win plan would be to seek outside help to get our foot back in the door, then seize the opportunity to demonstrate that our product was superior. We spoke to different partners—companies that already worked with the insurance company—to see who might help us secure another meeting. Two weeks later, we learned that one partner's Managing Director had spoken to the insurance company's CEO and asked them to reevaluate us. They agreed, but there was still a chance that our competition could do a better job.

With our partner's help, we made inroads and GREW our influence (the G in MAGIC). A couple of months later, we demonstrated our technology to the decision makers. We went in with how to solve their international financial reporting standards (IFRS) requirements (the Alignment) from end to end and how we fit in along with the benefits and differentiators. Shortly after that, they decided *not* to work with our competitor. They gave us the Named RFP, specifically asking for the brand name of the products we demoed. This is a classic case of late interception of a deal we didn't help create or shape. In this case the customer was educated over several months and our competitor was simply positioning their product. It is important when you are early in the sales cycle to make the end-to-end approach "reasonably" complex and 'risky' to get a new product vendor, especially in "net new" accounts. If that doesn't happen for whatever reason, you need to assume your ability to win is fifty-fifty and is dependent on how well the competition plays if it enters the fray late in the cycle. All we had to do was be quick and to the point while exposing certain aspects of the product that the previous vendor didn't share with the customer.

Our win probability went from less than 10 percent to 100 percent. The tables turned thanks to our partner's help. We knew if a partner organization the insurance company trusted suggested they reconsider our technology, we'd get our chance. They saw for themselves that our tech was better instead of assuming our competitor's was.

Our competitor should have focused on the customer's perception. We not only demonstrated superior technology, we also demonstrated a hunger for business, an intent to provide the best possible solution, and the skills to deliver it in a much better fashion.

WHAT EXACTLY IS A WIN PLAN?

At a fifty-thousand-foot level, a win plan is the ability to articulate with the MAGIC Lens what your chances are of winning. There are no prizes if you are a close second or in last place, so having a win plan charted out is a bread-and-butter activity daily. Use your own style to keep increasing the winning probability all the way to 100. Your starting position can range from zero percent all the way up. Competitors may start lower than your win probability or higher.

At a fifty-foot level a win plan is the specifics around these items that are covered by the MAGIC Lens. The fifty-thousand-foot view to the five-thousand-foot view to the fifty-foot view gives you a clearer context. The granularity will be discussed in this chapter.

On one occasion we were competing against a market leader against whom we had regularly won some deals and lost some. We were also a market leader depending on which territory you were looking at. The third competitor was a solid midmarket player with many customers and skills in the market. Against this player our company didn't have any win history in the region. It was a foregone conclusion that they were the price leader. The trouble was that this customer was not really large, so if that third vendor managed to demonstrate a good fit, they would win on price alone.

My first take was that we had no more than a 20 percent win probability, so how could we plan to increase our chances? The customers had a two-step process, and only two competitors made it to the final round. All we had to do was not lose in the initial round. We succeeded, and as feared the market leader was eliminated. That meant technically they qualified and price-wise we knew we couldn't go lower. Plus, we never encountered this vendor in the market as we played in different segments, so we couldn't draw on our experience with them. I put our chances at 30 percent. How do we increase those odds? With no clear lead on technology and an expected gap in price, the only way for us was to ensure a "frictionless" engagement with the evaluation team and be the vendor the evaluation team liked. I instructed our team of nine that for every objection we must either **convince** the customer **or be convinced**. I was looking for evidence that the customer was guiding us, and we saw it consistently over the next three days of intense negotiations. When a customer helps us, the win probability is closer to 65 percent in my mind. We were not driving for a win outright. We were driving to get a first right of refusal. That really means a one-on-one negotiation, which was exactly how it turned out. In my experience those extra final terms are usually negotiable, and it's the skill that determines how close you get to those conditions.

Sure enough, this telecom helped us win their business, and if two products are doing the job, then it's the people who make a difference. In this case, it was the people on my team. The telco gave us the first right to refuse, and of course we didn't. The $5 million deal was ours after eight days of intense negotiations.

How you do something is just as important as *what* you do. No two deals are the same, and no two win plans are the same. So, now that you know what to include in a win plan, you need to know how to approach each deal. To do that, we need parameters. Here are yours.

THE TWELVE FILTERS (OR LENSES) OF A WIN PLAN

What is the thought process for writing a win plan? The first thing to keep in mind is all the variables. Most salespeople and their managers think win planning is a chore, so they don't do it. This leads to "spray and pray" prospecting, selling, and objection handling. The odds of any one deal closing is so low that they have to try everything in hopes of something working.

Sales training sessions tend to focus on prescriptive questions that are more like memorizing a script to the tee. How many sellers can memorize a script? Can you? Especially if the scripts mean some dozens of questions for every aspect of the deal? If you are one of the salespeople who can memorize the script and ask all questions, then great. Keep doing it, if it's working for you. You also need to take additional care that those questions don't sound like a reporter interviewing someone. Another thing with these scripted questions is that not everyone appreciates the same question in the same tone. In some countries or situations you need to develop other means of asking them. If you have attended a sales training program and thought that these questions can't be asked directly, you know what I mean. The bottom line is that you need to be sensitive to and smart about the context of your sales opportunity, so I decided to approach it in a way that helps you realize what outcome you need, like a small piece of a bigger puzzle.

You know what the variables are that show up in a win plan, so the question is how do you assess them? How do you build a win plan that is simple enough for different team members to understand, align themselves to, and do their part to achieve? After all, a win plan is not just for you; it's for both companies, seller and buyer.

I've created a list of twelve different Filters for you to include all the essentials (and only those essentials) in your win plan. I've also used two of those twelve parameters as *x* and *y*-axes to help you visualize those parameters as a quadrant. You probably took algebra at some point in your schooling, so you should be familiar with the *x*-axis (horizontal) and the *y*-axis (vertical). Like everything else in this book, the twelve parameters cover about 80 percent of scenarios you'll ever encounter. The twelve parameters to include in every win plan are:

1. Your install base: Is the account a large install base, a competition stronghold, or new? (Defence, Offense and Net New)

2. Customer market position: Is this account a market leader or follower?
3. Customer behavior: Early adopter or laggard?
4. Who created this opportunity?
5. How is your (or your company's) relationship with the account? Great? Neutral? Hostile?
6. What is the customer buying? Full picture. Your products alone or many others?
7. Your market position globally: Leader? Niche? New entrant?
8. Your market position in the country: Leader? Niche? New entrant?
9. Your solution category: Mature? Fast growing?
10. Skill and will of your team?
11. Current playing form? (timeline pressure, etc)
12. Competition: Known? Unknown?

What is the right thing to do given these twelve lenses? The answer depends on your situational awareness of the deal. In this book, I've given you several examples of unusual wins in unusual circumstances. This doesn't mean that what worked for one situation will work exactly the same way in another specific context. That's why, instead of giving you a standard win plan template, I'm giving you a system—the twelve parameters—to help you assemble the right things to do in each opportunity.

The approach you're about to learn was created from the ground up to adapt well to a wide range of situations. It will put forth words that you've been struggling to find and help provide clarity of context. It has been my "red pill," a truth that permanently alters the way I view my profession. Once you learn the twelve parameters, you see them in play in every deal. What pleasantly surprised me is how my team, company, and partners have responded to the twelve parameters. This approach to win planning provides clarity in a way that enables everyone on the team to know how to contribute. As a result, my support teams, like partner managers, presales, and architects, have much higher win ratios in my territory than in others. That means they always prioritize my work first. That in turn has led me to be more effective, so it's a virtuous cycle over time, every time.

WIN PLAN PARAMETER 1: CUSTOMER'S MARKET POSITION

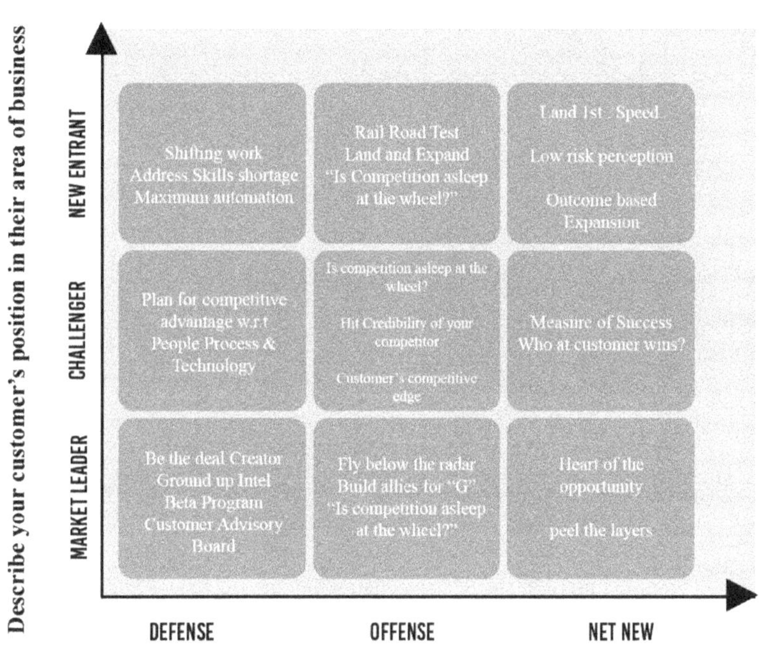

Describe this account to your company based on revenue, solution footprint and duration of time

The *x*-axis categorizes customers into Defense, Offense, and Net New customers.

The *y*-axis categorizes the customers by *their* market position. Are they a market leader, challenger, or new entrant in the market? Based on how each deal stacks up, The 3x3 matrix provides you with a directionally accurate suggestion you need to follow in the sales cycle. If your defense account is a market leader, you are better off planning to be the deal creator, with ground up Intel on the customer's business problem. The best way to get that intel is to have customers advise you on what to fix in your regular monthly/quarterly reviews. Customer Advisory boards or Beta programs are a couple of options to make that work.

WIN PLAN PARAMETER 2: WHO CREATED THE OPPORTUNITY?

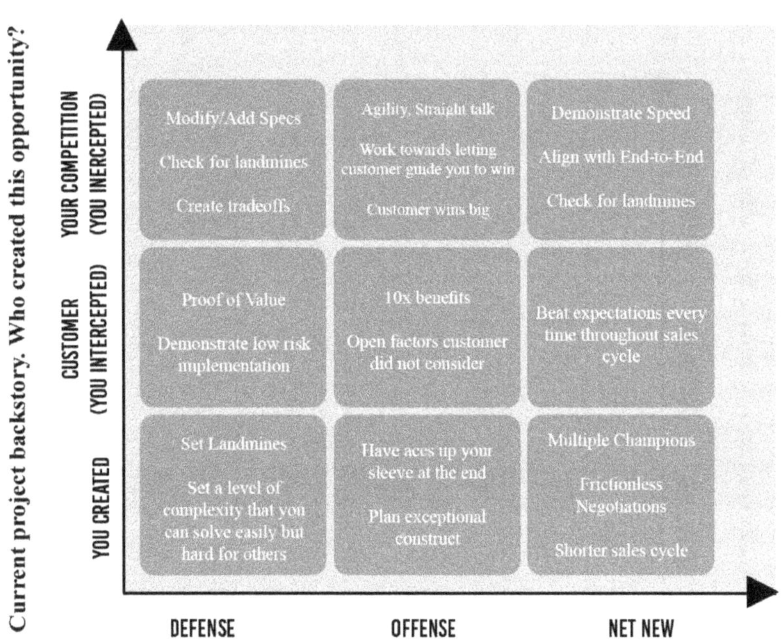

Describe this account to your company based on revenue, solution footprint and duration of time

The *y*-axis looks for who created the opportunity. If you get the opportunity to create one, it's best to set some landmines. When you create the opportunity you plan it in such a way that any new entrant will not demonstrate the same functionality quicker than you. Landmines are the moving parts that necessarily make it hard to overcome at the last minute. If your competition makes the opportunity, you cannot go down that same path. You must depend on allies such as partners and areas of weak alignment between customer outcomes and your competition's functionality. If the customer created the opportunity, it's likely a neutral ground. Sometimes it can be a case of all three aspects, in which case it becomes easier to put it through.

Once you identify the project's backstory, you can pick the approach that suits you best. Pick one square or a combination of them. If they're next to each other, don't worry too much about the border. Just be mindful that you could look at four of the squares together if you like.

WIN PLAN PARAMETER 3: CUSTOMER MINDSET

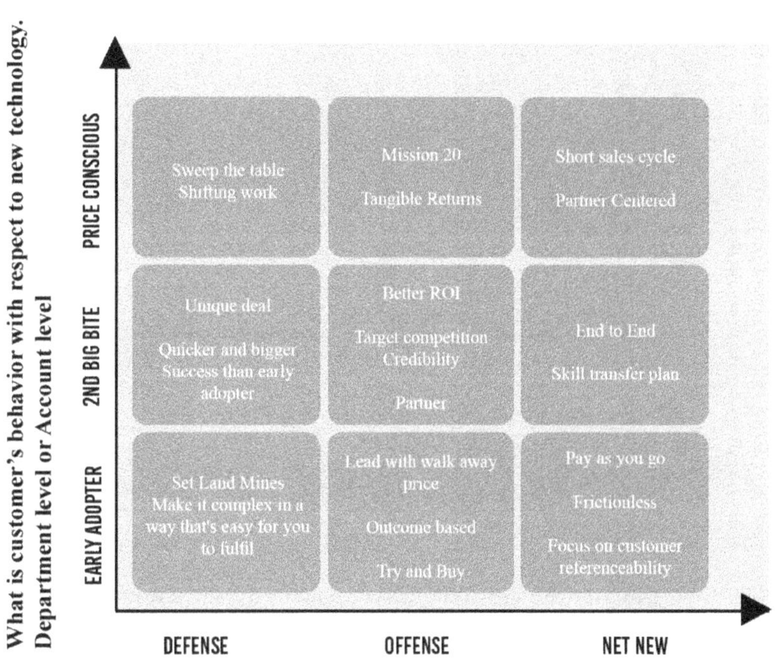

Describe this account to your company based on revenue, solution footprint and duration of time

Even if the customer is a major user of your products or services, their mindset can be used to guide your plan. Many companies are conservative while others are aggressive in adopting new technology. There are leaders, and there are followers. If they're a distant follower, perhaps they want lower costs. If your product is cutting edge, of course you want to focus on the clients with a thought leadership mindset. But if you're selling an old technology, take a very different approach to be successful.

WIN PLAN PARAMETER 4: RELATIONSHIP STATUS

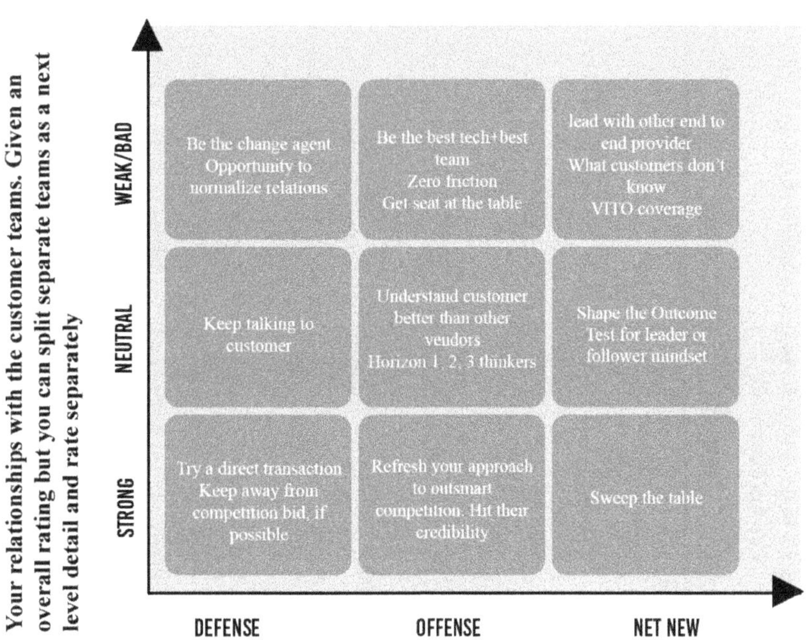

Describe this account to your company based on revenue, solution footprint and duration of time

This is a self awareness check about where you enter the buying cycle and what's your influence to start with. Be mindful about whether or not your relationships are strong. Even defense accounts, customers who already work with you, can have bad relations with your employer even to the point of legal battles. With such an account, one mistake is absolutely deal-ending. This is different from competitive opportunities where you don't have any relation, but they may not like your technology. Every relationship is different. Differentiate, understand the problems, and plan your approach accordingly.

WIN PLAN PARAMETER 5: WHAT IS THE CUSTOMER BUYING?

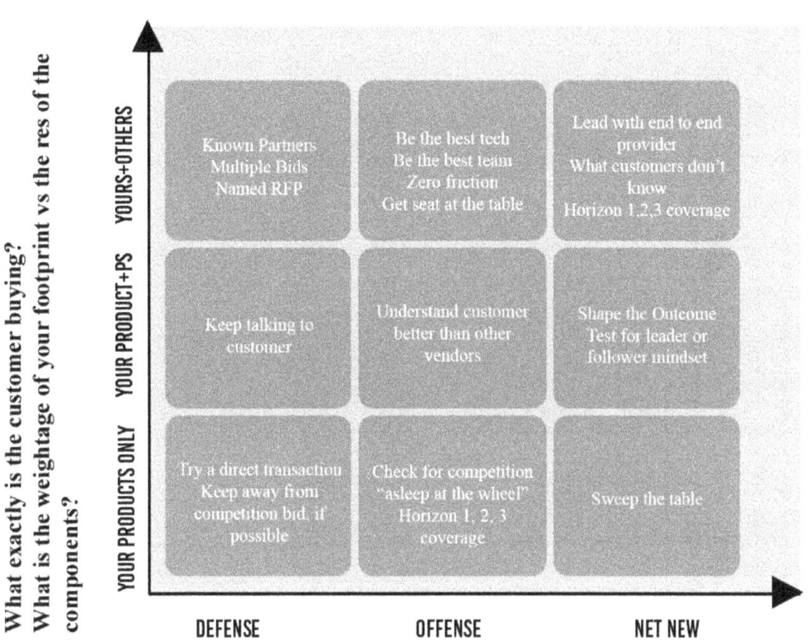

Describe this account to your company based on revenue, solution footprint and duration of time

What is the customer buying? Are they buying just your product? Are they buying services from your company? Or are they buying components from you and assembling them with others into a much bigger solution? How much of the solution you're responsible for determines how to approach the deal. If 90 percent of the solution is yours or if yours is the only product, of course you're in charge. But what if your component is only one part among many? That changes how you bid, how you interact, how you do everything. It's tough because what wins the deal could be playing a price game, depending on competent partners, or differentiating your features such as speed, functionality, or performance.

WIN PLAN PARAMETER 6: VENDOR MARKET POSITION GLOBALLY

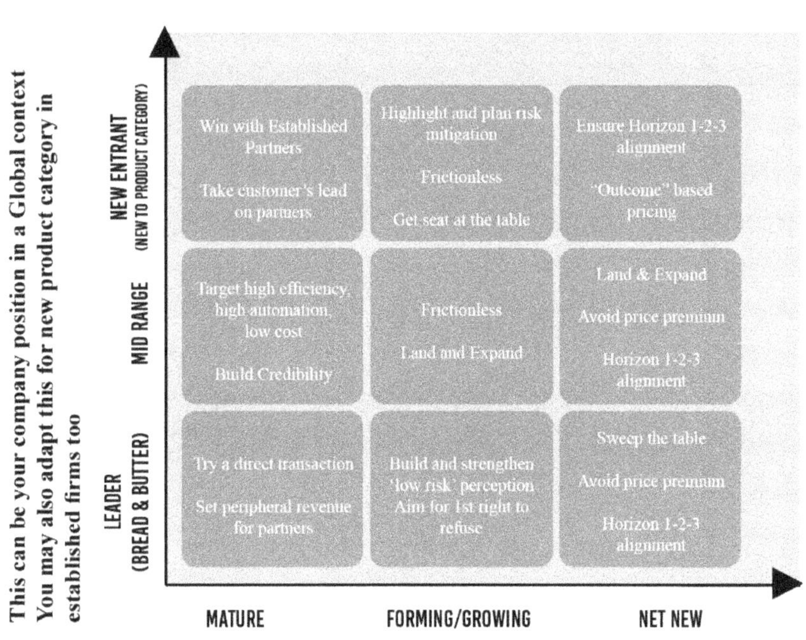

Describe your solution category.

The sixth parameter's *x*-axis differs from that of the previous five. Your company is either mature, fast-growing, or net new. If you're in a mature category like ERP systems, you've been around a lot longer than other emerging technologies for instance. I'm not suggesting there are no latest-and-greatest ERPs, but the setup, cycle, costs, and returns are very well known. If your solution category is forming like the artificial intelligence and machine learning space nowadays, your win plan will be different. Even more different still are net-new solutions such as nanorobotics, an emerging technology using machines at or near the scale of a nanometer.

Based on your solution category, move up the *y*-axis. Your own company's market position varies by deal. You may be the biggest hardware or the most-trusted software vendor. If you're the world's bread-and-butter vendor in your category, you will sell very differently from midrange players or new entrants. If you're either of those, consult your square and follow the plan.

New solution categories, which are a common sight these days with mushrooming startups (there are also large companies building new ones), catch the attention of Senior Executives. They're the right people to engage for such deals either directly or with help of (usually) consulting partners. But be careful with the qualification process. In the early days of a newly forming category for big data in Asia, we made quick progress with the help of a partner to meet the CEO of a large regional bank. We had all the right messaging, aligned with the problems he wanted to solve, and demonstrated our capability. He then invited three of his lieutenants (from three different teams) to work together and get this done. Our excitement was short-lived. The bank was not ready in terms of structure with roles responsible for specific outcomes. Meaning Horizon 3 (CEO, in this case) supported but the company didn't have Horizon 3 and Horizon 1 people in place with the necessary roles for those specific outcomes. The project required us to work with those three teams separately, each with their own KPIs. No lieutenant wanted to be responsible for the full outcome beyond their own KPIs. It took us eighteen months after we got explicit support from the CEO after Horizon 2 and 1 were aligned.

WIN PLAN PARAMETER 7 VENDOR MARKET POSITION IN COUNTRY

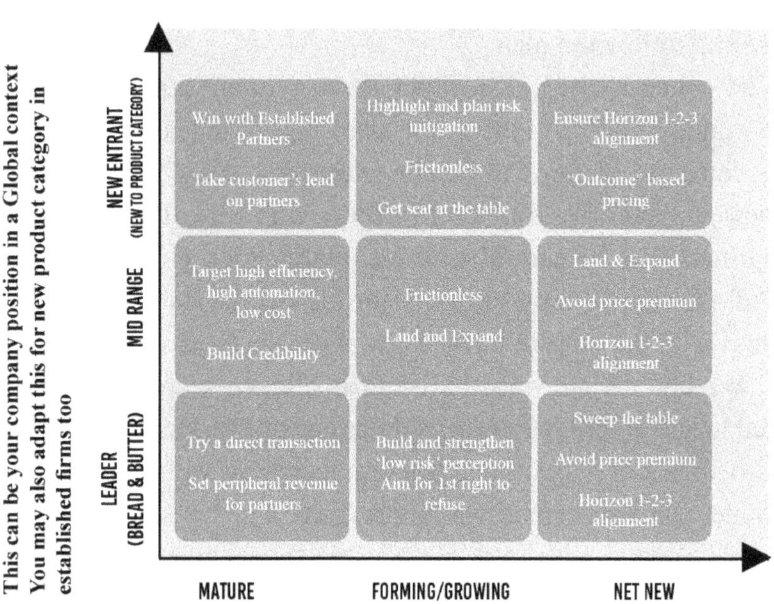

Describe your solution category. Remember that new solutions in an existing category are different from new solution categories. This picture refers to new solution categories. In country dynamics take precedence over global position.

This parameter's *x*- and *y*-axes are the same as in the previous parameter, so let's talk about what makes this one different. In some cases, you may be a global leader, but you should adapt to country dynamics. Identify whether you're a market leader or a follower in your region. It may be that you are the biggest global player but may not have any install base in a specific country. Sell accordingly. In one emerging country the core banking system a customer selected was a company with more than 80 percent of their underlying platform on our competition. Only a handful of installations worldwide used our platform for that application even though our product was the world leader. Our win plan was to expose their lack of due diligence for the platform and highlight the risks around accountability in the event of downtime. The competition was not there to counter as they didn't think we were in the fray to dislodge them. They must have not followed magical selling since the application vendor was positioning their platform anyway. The reality was when the customers asked for our platform instead of the competition's, the application vendor didn't object. These are fertile grounds to find competition asleep at the wheel and result in easy pickings.

WIN PLAN PARAMETER 8: STRENGTHS OF YOUR TEAM

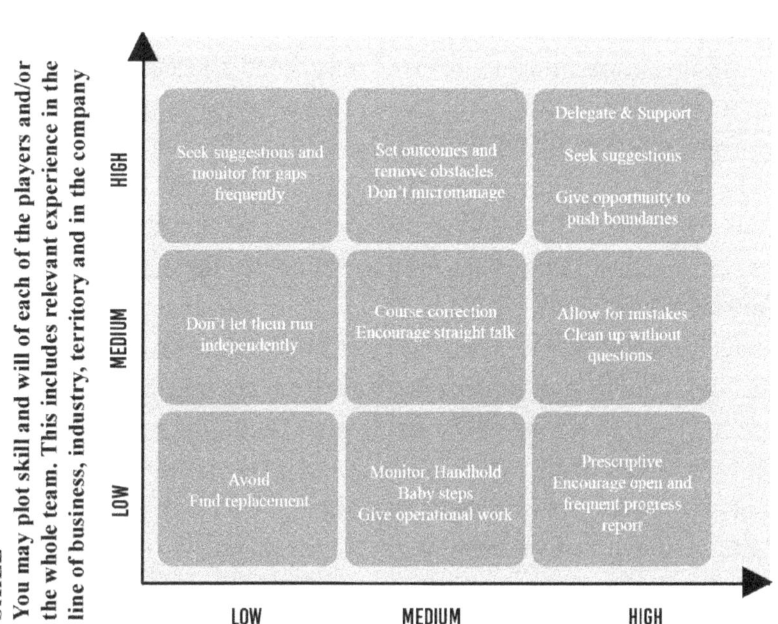

WILL of the team member collectively or for each team. Teams with clarity on deal context and on win plan perform better than otherwise, even if the team members are the same.

RESOURCE ALLOCATIOWN IS A SKILL

For this you need to think like an investor. The eighth parameter's *x*-axis is will while the *y*-axis is skill. "Will" refers to desire to prove yourself as well as your energy to see the deal through to the sale no matter what. "Skills" refers to not just product knowledge and sales competence but also to experience with the account or similar ones. If you or your team do not have the will or the skill, consider partners, including other people within your company. The idea is to ensure you have the right skills that are aligned with the job at hand. My favorite is to look for volunteers for any aspect of the plan. Let's say for convincing a certain individual or department you identify a specific scope. Who would be willing to do this? Or who is the best person to do this? Consider anyone who is willing to volunteer. It means they know the work and are committed to it. You only need to decide whether that person needs supervision while doing that activity or if they can be left alone. Take a soccer match for instance. If you have a forward like Lionel Messi on your team, you don't want to put him in a goalkeeping position. The joke will be on you. While that may look absurd in reality, I have seen sales situations where this is exactly the case.

One such case involved Rizwan. I was new to that territory, and my first meeting with the team was a quarterly business review. This came with a lot of team dynamics that surprised anyone looking from the outside in. There was one particularly important deal in the government sector, and in the deal review the head of that country's sales and the account team was not confident that the deal would happen. The discussion was a kind of informal win plan discussion—informal because I was still learning what was going on with the team and the market before setting the cadence structure. While everyone in sales was not confident and therefore not forecasting, Rizwan, who was a presales guy, had a totally opposite view. He said he knew a key officer at the customer and went on to explain why the deal could be pulled off. For the next half an hour there were views and counterviews, and my final decision was that Rizwan would front end the deal and the rest would support him. Rizwan had a technically strong background, so he didn't need help. I offered to support him in other aspects to cover the deal while he set it up with the specific people he knew personally. During this time he would report to me directly with others participating in their respective roles. I gave him this responsibility and told everyone that if it went south I'd be held accountable. I also made sure there were no other deals running short of resources, as it may turn out that Rizwan's spending time on that deal was really affecting other sales. Once this risk was mitigated, it was a matter of executing the sales cycle well.

Over the next seven months, Rizwan was very diligent and disciplined about bringing the deal home. Rizwan loved the fact that his idea was valued, and even in a lot of opposition he got a chance to demonstrate the skill to bring it in. The others saw a decision happen in a very objective manner with plans to mitigate the risk and how it went on to close. People tend to learn based on what they see, not what they are told. Once you go through a journey like that, you will see a level of trust gets stronger and the team starts winning more. We grew that business by 50 percent within eighteen months.

I have seen this repeatedly in other roles. My role has been as a regional manager for a couple of decades now, and other regional managers and specialists keep flying down to meet customers whereas the local sales team doesn't tend to travel much by comparison. In a way it is considered that regional guys are more highflying, but then many of the regional guys are not as smart as many local account managers. I used to test this out, and most of the time the reps would say—usually in one-on-one meetings—that they could do a better job than the regional folks. The way out of this was to allow them to volunteer, agree on a plan and leave them alone or monitor them without doing the work yourself. You will be surprised how many people will develop the skill and will to succeed almost like magic. It happened with another sales rep much more junior than I, and after a few meetings I had with him, he actually told me that he could do a better job than me in a couple of meetings and asked me to leave it to him. Guess what? He was right.

He was happy that he was able to make a difference, and I was happy to support him.

WIN PLAN PARAMETER 9: CURRENT PLAYING FORM

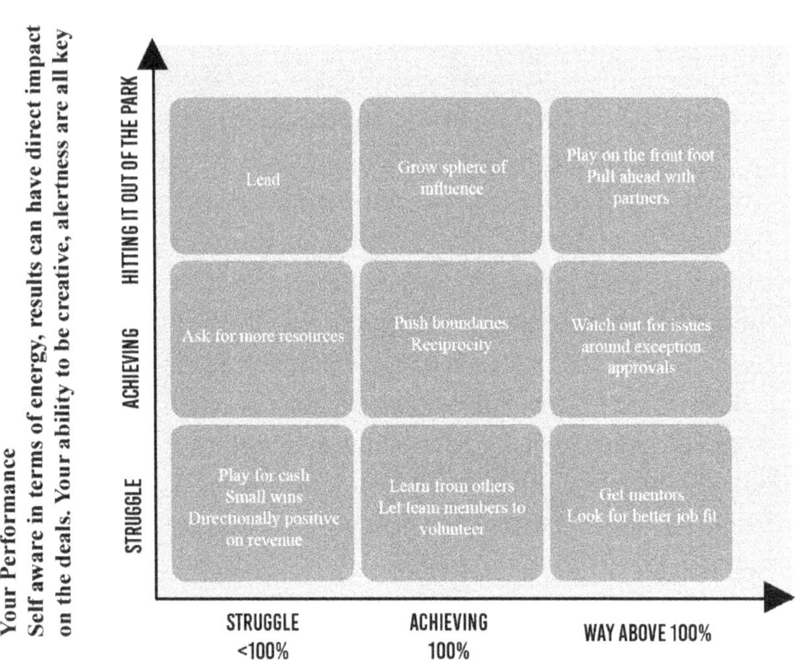

Company's current performance. There is always a situation when you might want to change jobs. For the purpose of this chart we are focusing on staying with the company, even if you are struggling

The ninth parameter's *x*-axis and *y*-axis are similar. Here the focus is on your company's performance—struggling, achieving, overachieving—as well as your individual performance. Have you been winning a lot this year? Losing a ton? You may be hitting it out of the park while the rest of the sales team struggles. Or vice versa, you're the only salesperson who cannot seem to close. Follow the axes to know the right way to get back into form or to build on your current performance and win even more.

WIN PLAN PARAMETER 10: COMPETITION

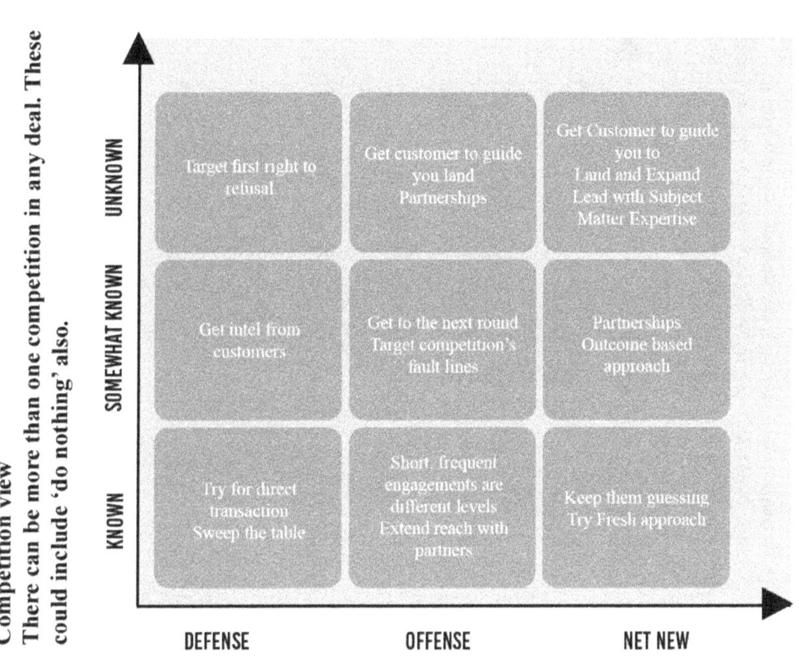

Your perspective of this account

The x-axis is back to defense, offense, and net new. With the y-axis, it's important to know what your competition looks like. Do you even know who they are? Are you generally familiar with their unique value proposition? Have you battled each other for a customer before? In more established markets, you're usually familiar with the opposition, whereas in less established segments you may not know or be familiar with them. This is most relatable for newcomers or territories where disruption is happening. As with all other parameters, look at the x-axis, move up to the y-axis, and follow the plan.

WIN PLAN PARAMETER 11: MARKET CONDITIONS

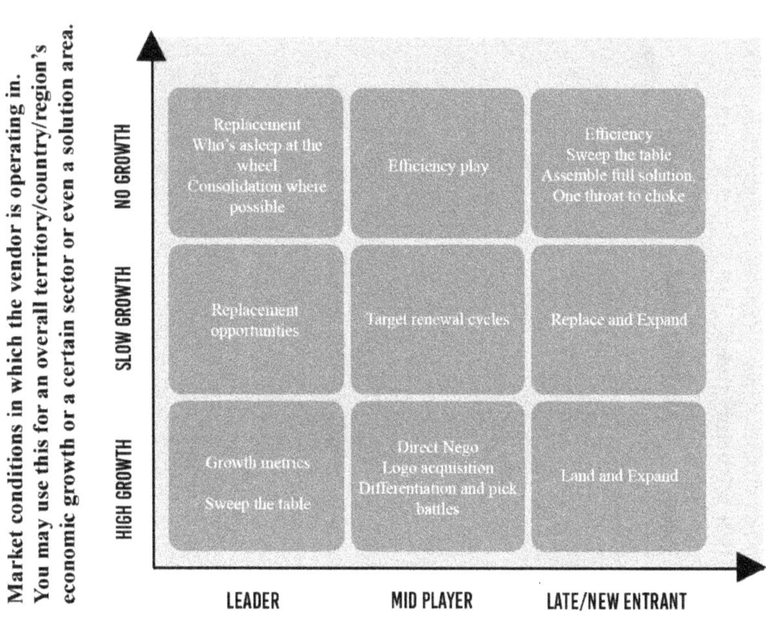

Don't forget about market conditions. COVID-19 has changed a lot of markets. To be in the tourism industry in spring 2020 was not a good thing. Leaders have a lot more to lose, so their ability to defend their position is important. If market conditions allow for high growth, take calculated risks. Accelerate flexible pricing models. Again, follow the *x*- and *y*-axes to success.

WIN PLAN PARAMETER 12: TERRAIN KNOWLEDGE

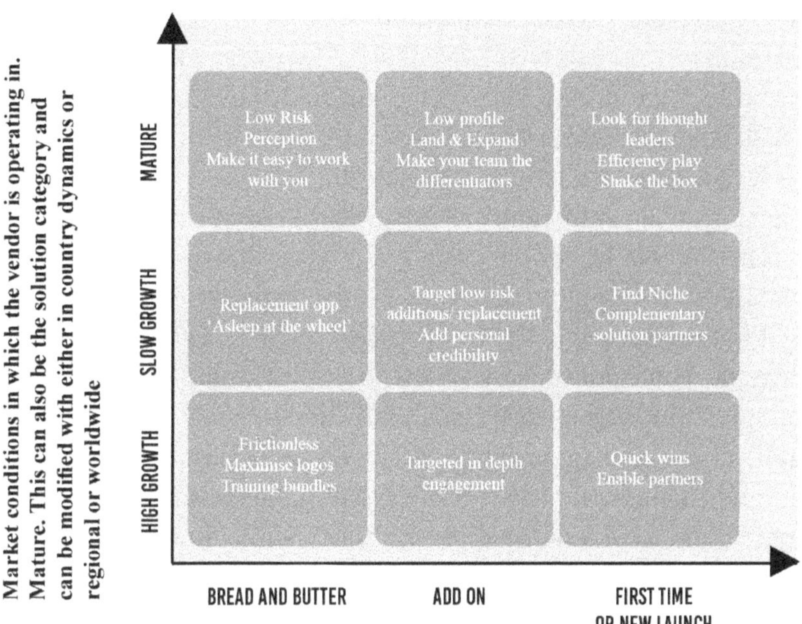

Vendor status in the market for the product/solution being sold. These are typically established players expanding with adjacent solutions they are not known for. In other words, selling weaker or less known products of an established company

When I say "terrain knowledge," I mean the sense that you may be known for your product, you may not be known for the product, or you may be competing against another vendor's bread-and-butter product. When you're taking responsibility for your own area of knowledge, your own bread-and-butter product or solution, base your win plan on that. It's the other way around if it's somebody else's bread-and-butter against your new product nobody knows about. All companies start with a core offering and gradually expand into related fields over time. Companies that have go-to-market figured out for such expansions like to have effective specialists for that product line. These are the ones who succeed. I've seen many startups with a core offering. Instead of scaling that core product, they develop the next one. Usually, this is based on feedback from customers who adopted the core product early. It looks lucrative but is often a trap. Hardly any thinking through is done in such cases because the adjacent expansion opportunity may face cutthroat competition against established vendors. Then you wonder why it's not selling. Keep that terrain knowledge in mind. Find a niche, and don't ever just take the competition head-on, especially if you're working with a product that isn't your bread-and-butter.

These twelve parameters are lenses that give you different perspectives. They're really twelve decision points. You may use additional combinations to adjust your win probability and provide clarity of thought to the question: "What is the right thing to do?"

Why do you think this approach to win planning works? In sales, far too many times, we work from a ten-foot view, close to the ground. All trainers and managers simply focus on what is right in front of everyone and end up with the spraying and praying approach because they cannot tell you why what's happening is happening nor can they predict with accuracy what will happen next with any in-progress deal. To be effective, you need a ten-thousand-foot view and also a hundred-thousand-foot view. Your ability to switch among these three views at will for any deal automatically brings out your "smart sense" that makes you successful. After all, **smart people combine deep knowledge, good decision-making, and skills to succeed at whatever they do.**

I don't want to be prescriptive even though some people prefer it. Prescriptive advice is appropriate for one-to-one sessions when I'm consulting with a team to customize and propel their sales careers. The twelve "lenses" I've provided will help you change perspectives instantly so you can see the best angle with which to win. Then use your own unique execution style and company templates.

Have you heard the phrases "man with a plan" or "sounds like a plan"? Whenever I teach salespeople how to win deals with the twelve parameters, I always hear those phrases as they describe this approach. Mapping win plans well and consistently will get you the single most important attribute of a leader, which is having followers. If you are effective, there are enough people in the world to follow you. Leverage win planning to create a position of power and value for yourself.

If win plans for each of your deals look a bit different from the previous one, you are on track to improve your chances of winning that next deal.

TOOLS TO HELP WITH TACTICAL DECISION MAKING

When you are done evaluating your way through these 12 (+/-) lenses, you may find a way to move forward in your mind or finalize your next steps. For the sake of clarity, I'll give you one last 3x3 matrix that may be a simplified, suggested approach. The individual elements within it give you directional clarity on the best way to approach the deal in order to align your message internally and externally with partners and customers to help you win. In my experience this has a high ratio to win. You will notice that you may have heard or read about these wisdom modules before. I'm just assembling them for you to even consider having them as "memes." Don't be surprised if you end up using a variety of other aspects like interviews, planning, and so on.

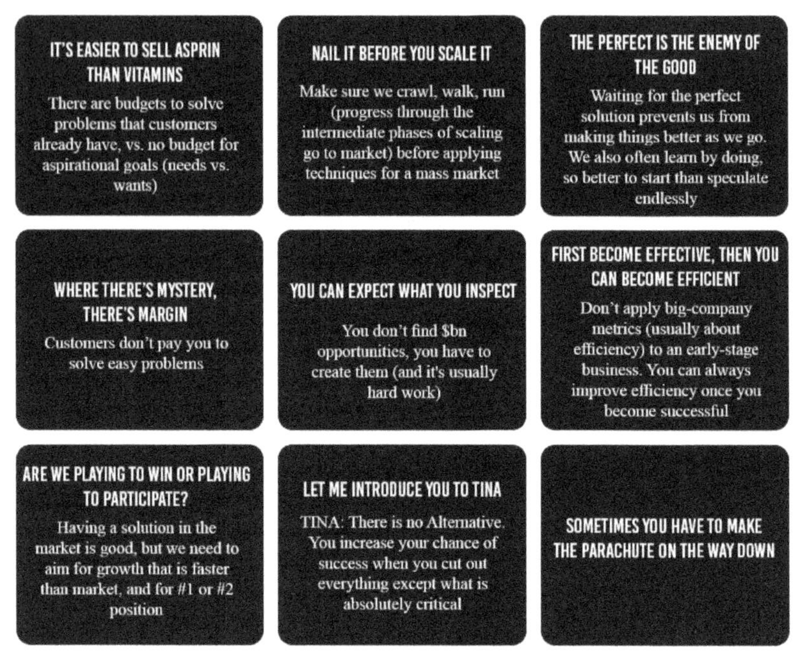

Figure 10 The 3x3 Sales Matrix.

This simple matrix will get you far. I would like to promise that it will get you all the way, but that is not always the case. In certain situations, more advadanced, nuanced advice may be required. Let's look at that next.

TROUBLESHOOTING YOUR WIN PLAN

In the same way that a GPS system updates the map based on traffic, construction, weather, and other conditions, win planning is done in real time. Updates to the plan, we call troubleshooting.

"Troubleshooting" involves a weekly forecast call where you get together with your team to discuss what everyone is doing to follow the win plan, what variables have changed, and what must be done to course-correct. Leading this call is the responsibility of whoever is the quarterback, be it the sales head, account manager, head of the region, or even the CEO.

Troubleshooting is most useful when the deal appears to be falling apart or at the very least the win probability is decreasing for whatever reason. Salespeople tend to say it's not their fault. The customer is back-pedaling the relationship because of pricing, the product is bad, or some other reason. So says the salesperson. These are never the real reasons. Promote straight talk with your team. You cannot guess why progress has stalled. That's a wasted discussion. When you meet to troubleshoot, refer back to the win plan. How must your approach change to improve your chance of winning the deal? Look forward, not backward.

Most often, troubleshooting takes the form of objection handling, so that's what we must cover next.

OBJECTION HANDLING FOR THE WIN

A casual Google search will show you dozens and dozens of objections listed on different websites. Some sites have over one hundred objections. I would caution you against anyone using those websites. The person writing simply didn't do enough selling. If you try to learn from them, you are only wasting your time and as a bonus losing confidence because the sale now seems too difficult.

Selling is an umbrella term used to describe the process of closing a deal, persuading people, convincing people. Negotiation is an important part of selling that is dealt with as an independent topic, which I agree with. The one absolutely critical underlying constant that doesn't receive enough attention is "objection handling." If you are good at objection handling, then you will find negotiation and in general selling easier. Sales is a rejection business, and even the ones that we close go through a truckload of objections. As you know the weakest link in a chain determines the strength of the chain. It's like strengthening your core to run faster, which strengthens the arms and legs. Strengthening your objection handling skills turns you into a strong negotiator or seller. In fact, smart negotiators have a way to nudge customers toward an objection that they can resolve in order to set up the deals for closure. Yet it's surprising that there is a lot of mediocre content out there on objection handling. Instead of memorizing dozens of questions and in the end turning out to be a machine, is there an easier, more effective way to improve your objection handling that's also fun? How about something that applies to our day-to-day life too instead of just our lives as salespeople?

When sales reps pursue a prospect, the first reaction nine times out of ten is, "I don't need this." Sales trainers teach us to accept this rejection, shake off the dust, and move on. That's motivational but helps nobody. Reject the concept of rejection. It's not you who is being qualified; you are doing the qualifying. Should you sell to this person or not? Customers appreciate salespeople with this attitude because they will not be disturbed if they're not a prospective buyer. If they're not suitable, thank them for their time and go. That leaves a good impression; you're smart enough to let them go.

In a negotiation, most salespeople go in with a warrior's mindset to battle through any objections. I go in knowing I will gain clarity on whether to go or not to go and to discover the most effective way forward. Objections are present in every deal, even transactional sales, but most salespeople do not follow a system to reach a mutually beneficial outcome. If or until a better one comes along, I recommend we use the objection handling techniques I've honed over the years. They work better than anything else I've tried. I classify them into three simple categories that anyone can understand and start practicing. They won't take much time to learn. What will need time and practice is how accurately you can classify any objection into one of these categories and follow the system to move your deal forward.

The first type of objection is the one based on **misinformation**. Misinformation occurs when the customer, a partner, or even your colleague or manager misinterpreted what was written or said in relation to the deal. Misinformation is not unique to sales; it lurks in every human interaction.

Handle misinformation by clarifying. Ask a question, clarify, apologize if you misread, then restate your point accurately and test your understanding. Once you do, stop talking. Far too many salespeople use every objection-piercing arrow in their arsenal on one customer concern. Be smart. You don't need to do anything else other than apologize where required and restate what the customer now understands to confirm you're on the same page. Then move on.

The second type of objection is **disinformation**, which happens when a customer objects internally or externally. Anyone who has a significant other, kids, friends, or colleagues has experienced this. Disinformation is when you hear a deliberate play to admit you have done something wrong. In business, this usually comes from a competitor, who told the same customer you're both pursuing to beware of your company. Other times, the source of disinformation is negative PR, a false story planted in the news media, or an antagonist who works within your customer's company but for whatever reason does not want to do business with you.

Most disinformation is misinformation. It's rare that someone is out to get you. Yes, it can happen, but it's more likely that someone the customer trusts misinformed them about you, your company, or your product, and they assume the information is true. You don't need to defend yourself or attack anyone. Do not let it escalate. Handle this the same way you handle a domestic argument. Let's say I walk into home, and my wife asks me how the movie was. What movie? She heard from a neighbor that I was seen at the movie theater. I tell my wife I wasn't. Before I let the situation get out of hand, I lay out the facts. I was giving a product demonstration at a restaurant near the theater and walked past the theater on my way out—a completely innocent and entirely true explanation. Tell your side of the story without emotion and offer proof as required. For example, let's say I produce the restaurant receipt.

That's how to handle the disinformation objection. In business, that means you offer a proof of concept so the customer can see for themselves. Other ways of doing this are third-party reports or benchmarks and even customer reference calls on how well your solution is helping them drive better outcomes. If there is a real disinformation campaign running against you, use your clients as references and use case studies that show how your product works and how you've responded when issues came up. If you correctly identify an objection to be based on disinformation, don't try to explain yourself and clarify. It will be counterproductive. I would urge you not to use a combination of explaining and showing proof. It dilutes and weakens your position. Nothing more, nothing less. Just be sure to validate that you have overcome that objection and move on.

The third objection is a **genuine drawback.** These happen when, for example, the customer says you don't offer a particular feature. If that's true, it's a genuine drawback. What do you do with a genuine drawback? Do not explain yourself or offer proof. You won't have proof, of course, and if you have proof, it means it's not a genuine drawback. The best response is to acknowledge the drawback and explain it in context so it is not blown out of proportion. The idea is to educate the customer on why the issue they perceive is not that serious. That means *you* must first understand what features drive which business outcomes.

As you listen for genuine drawbacks, keep in mind that customers are often irrational. I don't mean to be derogatory. There are times they veer off-topic or lose objectivity. You need to step in and change the course, especially if customers do not seem to be not making an objective decision about the project.

Let me give you an example. Years ago, we were helping a bank with their online transaction processing application. They needed an end-to-end solution, which they referred to as a "lock-in," which tipped me off to fear, uncertainty, and doubt about being at the mercy of a single vendor with a comprehensive solution. That perception was the genuine drawback, that if the one vendor they chose would be unable to troubleshoot a problem, potentially millions of banking customers would be unable to use their debit cards, make online wire transfers, or process other transactions.

We acknowledged that it was one perception but first explained the end-to-end benefit of having a fully integrated, tested system—reminding the client of the 90-plus percent of good things that came from having such a system—and discussed how we lower the risk of us misusing or underperforming in that situation. We put in price-hold guarantees for five years so that new people wouldn't take undue advantage of the customer and provided provisions to handholding, escalation paths, and executive sponsors for strengthening the relationship further.

That's when the bank's CIO, an otherwise sensible guy, started comparing chips. Whose computer chips were better? In the past, vendor decisions were made based on who could do more calculations per second. The CIO said that the chip we used for all applications seemed to be 20 percent less efficient than the alternative. That comparison was based on TPC benchmarks and therefore a genuine drawback.

Enter course correction by acknowledging and highlighting the overall benefit far outweighs the drawback identified.

"We can compare the chips over and over, but what happens when the system goes down?" I asked. "The system can always go down either at a storage level, operating systems level, or in any other component regardless of what chip you are using. When something goes wrong, your job as the CIO is to make sure the bank opens the next day with everything up and running. It doesn't matter whether the chip is five or ten percent less efficient if you cannot troubleshoot because you have four or five different vendors to sit down and work with because you did not choose a single end-to-end solution. Let's leave the chip calculations to someone else," I added. "Every chip gets improved over time, including ours. Which is more important, a difference in chip speed that banking customers will never notice or the certainty that when they use your online banking services, their transactions will process?"

"Thank you for reminding me of my job," the CIO said. He sounded genuine. "I appreciate you bringing us back to that. My real interest is efficiently running this institution, not searching for where the trouble is, which component is going wrong, and why I need to talk to five different vendors to try to fix certain issues."

We won the deal. Anyone who sells end-to-end software to banks knows the huge margin associated with such a sale. If I had not dealt with the customer's irrationality and instead tried to handle a genuine drawback that it turned out didn't really matter, I doubt we would have closed the deal.

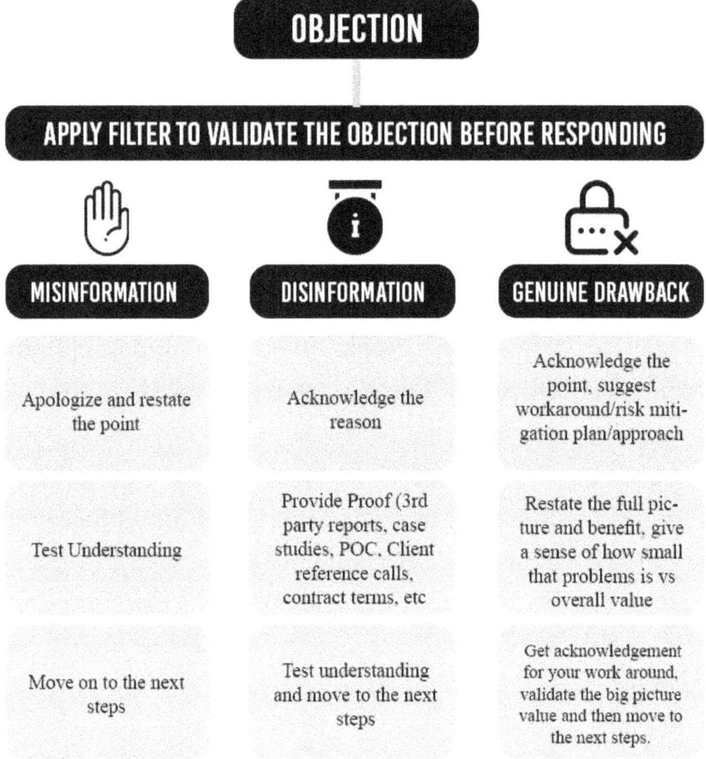

Figure 11 Objection Validation Process.

If you can identify and classify an objection as one of these three types, you will solve the problem quicker than any competitor. Some salespeople do this instinctively, but anyone can. All you have to do is listen. The nuance is in the classification. When you're selling a run-of-the-mill product in a transactional sale, a question about a missing feature could be misinformation. If you're selling a complex enterprise solution that will take months to roll out, the missing feature could be a genuine drawback you must account for. The more you identify, classify, and respond to objections, the better you will get at this. There will always be mistakes, but you will be able to spot where you made them and course-correct for next time.

Knowing there are three and only three types of objections is ridiculously useful because your objection handling technique is a repeatable, predictable system even when you get an objection you have not heard before. You're able to think on your feet and handle objections almost effortlessly. Most salespeople go into sales conversations hoping they get an objection they've gotten before and were able to answer. So much of sales training is basically, "Well, it worked before when we did it this way, so you should do it that way too." In no relationship in real life does that work. If you're at the bar making a new friend, you don't think of what you said to the last friend you made and repeat it. Every person and every situation is different.

Whatever objection you get, another tip I can give you is to retain control of your mental state. Thomas Anthony Harris's 1967 book *I'm OK—You're OK* is about the three states of mind—parent, child, adult. In the parent state of mind, people use phrases typically associated with being a parent. You *should* do this. The customer *should* be using that. Your boss *could* stop being that way. The child state of mind is complaining about everything without a sense of power to change anything. Meanwhile, the adult state of mind is logical, cool, and objective, trying to get on with life and not taking anything personally.

The three states of mind are useful to avoid friction when you're handling objections. Friction destroys deals. The more you're unable to see eye to eye with the customer, the less likely you are to win the deal.

I won't expand on the three states of mind because there are books written on it. Go read those if you feel inspired. The point of the mind states, for our purposes, is to understand objections and shift states as needed. As you're listening to the customer's objections, note which of the three states of mind they're in as they speak. When someone is in a parental state of mind, it's better to respond with a child state of mind because a parent and child conversation tends to be smoother. If you're in the parent's state of mind, and you're talking to someone in a parent state of mind, it won't work no matter what you say. It's your job to shift quickly according to the customer's state of mind. The one exception is if you are both in the adult state of mind. If you're both there, stay there.

Some call this manipulation or persuasion. I call it frictionless objection handling. Whenever you encounter a tough objection, the three states of mind are an additional tool you can use to fix the situation, appear calm and collected, and take the conversation where it needs to go.

ADVANCED DEAL TROUBLESHOOTING

Not every reason a deal stalls is considered an objection. Occasionally, the customer does not proceed toward closing the way we would like because they never had buyer's intent in the first place, the deal size is too small, or you've made an unforced error. The following sections show you what to do if you ever find yourself in those situations. Be forewarned—sometimes, winning means moving on to a deal you have a better chance of closing. You'll see what I mean.

KNOW WHEN TO LEAVE THE TABLE

There are situations where customers don't want to buy, they just want free consulting. The deal is a waste of time because it will probably never close. I experienced this with one of the largest banks in the country. After a couple of customer meetings, I had still not met the person who would sign off on the supposed deal. I'm a big believer in meeting the final decision maker to discuss why they would buy and what their time frame is. When I finally had that conversation, I realized it was all smoke and mirrors. The bank had made us believe they were interested so they could get free consulting. I took the deal out of my pipeline so we could save time and effort and pursue other deals. We did follow up in a year's time, but the deal still didn't go anywhere.

It's important to know when to offer the customer your subject-matter expertise. It's also important to call out any customer who does not intend to buy but wants some free support. This is common, yet most people let themselves be taken advantage of. They offer advice, let the conversations continue, keep the deal in the pipeline, and make excuses. No, the market isn't the problem. No, it's not a territory or language issue. No, you don't need to be more patient. Be objective. See this for what it is—theft. The customer is stealing your time, energy, and expertise, and you're letting it happen. Don't.

DON'T SHOOT YOURSELF IN THE FOOT

I love telling hunky-dory we-won stories, but that's not the case all the time. Just as every salesperson experiences winning, they also experience losing. The same goes for me. The only difference is our win ratio has been significantly higher than the average company's and much higher than many others in certain product lines and new businesses. It's important to know that no deal comes with a 100 percent guarantee. A win plan well drawn and well followed will bring you as close as possible to certain victory, but there is no certainty. Even so, a win plan will allow you to predict the outcome—deal or no deal?—sooner than any competitor so you can withdraw, regroup, and pursue the next deal.

A notable deal I lost was a big data deal three years ago from the time of this writing. I was leading a newly formed big data team to pursue prospects in that territory. I still had a lot to learn. I assigned a few account sales specialists to an opportunity with the largest telecommunications company in that country. We scouted for partners to collaborate with on the deal. We did not want to pitch the telecom company our solution directly because there was bad blood between their management and ours over a previous deal. Still, we knew we could work with them. When we were unable to find a partner to bid on our behalf, we went with direct engagement. I brought the telecom's RFP back to our consulting, product, hardware, and software teams, wrote our proposal, and pitched the customer. Typically before the proposal, we communicate with the customer directly to understand their needs and to show them we understand them. I was reluctant to do so on this deal because of the feud. Basically, we were pitching blind, and I knew it.

The approach I took I call "TINA": There Is No Alternative. I was prepared to win or lose with our first proposal. When we presented the proposal, the engagement went surprisingly well. Around thirty people attended from both sides, including those in technical, legal, compliance, and business operations. After the meeting, word got back to me that, despite the bad blood between our companies' leadership, the decision makers for the big data deal very much liked our teams. They took our proposal seriously and verbally agreed. The lesson I took from this is that customers want the best solution even if there are certain aspects about the vendor they may not like.

Meanwhile, the telecom company did their due diligence, accepting proposals from six other bidders for the deal before agreeing to ours in writing. We were clearly the frontrunner. Subsequent meetings between our teams gave us additional insight and shored up our understanding of their situation. Soon, we were one of only two vendors considered for the big data project.

One of the telecom company's decision makers came to us and told us our proposal was strong. So was the other bidder's, but they were strong in different ways. The customer was comfortable with our product line, but the other vendor offered a local partnership for on-site implementation with people who spoke the local language. Our product, their implementation. Like an idiot, I assumed they were considering buying the local vendor's product, which competed directly with ours. In my mind, the negotiation was about taking our products away from the deal and replacing them with the local implementation partner's. Since it was a complicated process, I thought it was possible that the customer believed the local partner offered something we did not. Negotiation mode kicked in, yet there was nothing to negotiate; I handled an objection that existed only in my head.

When the conversation turned for the worst, I figured office politics were involved. Perhaps managers within their company who didn't like ours had outsmarted us. That was not the case. I had screwed up the deal. My tactical error was to assume there was no alternative to the proposal we first submitted. In every deal since, I've kept this lesson top of mind: it's important to be flexible. Customers have their own preferences, their own licenses, their own understanding of technology, their own understanding of the region. Unless we're open to their needs, we cannot win the deal.

When the telecom company dropped our proposal and found another vendor to replace us to work with their local implementation partner, one of my internal contacts told me they really had wanted our product. We lost the deal because of me. It would have been worth $3 million. The mistakes I made with that deal, I will never repeat.

A year later, I lost another deal for an unrelated silly error I should have seen coming. It was a big data analytics deal in which we met the president, chief technology officer (CTO), and head of business. We followed our win plan up to and including proof of concept. As with any complex deal, there were hiccups along the way, but toward the end, the customer thought our proposal looked good. When they asked to negotiate terms, we had a sense of the deal value they would accept, and we were in position to meet them.

Once we handled their pricing objection, the customer asked us to compare our product with some other product that wasn't well known or active in the market. Instead of asking why they were considering the inferior product, I assumed their intent—they just wanted to put pressure on us. Nobody else had offered a product demonstration or proof, so what else could have been their reasoning?

One week before the customer's final decision date, they hired someone to manage the big data architecture setup. It turned out that this person had worked with our competitor's product, which was why that objection came up in the first place.

The day before the deadline, we met with the customer. I emphasized our superiority on every front—20 percent more power, 20 percent more software, 20 percent more results than our competitor. The deal value at that point was $2 million over the first two years with a 20 percent margin, and I had a little room to lower the price. Instead of issuing our final, lowest-possible "walkaway" price, I decided to wait one more day. Once they compared our superior offer to our obviously inferior competition, they would return to us with a counteroffer at a price reduction I knew we could accept.

The counteroffer never came. The customer considered the numbers I gave them in our last meeting to be our *final* offer. They called neither our side nor our competitor's for revised pricing. They awarded the contract to the lower of the two bidders, which wasn't us. The only reason they brought another vendor into the discussion in the first place was because their new employee told them the other product came with a better price. If I had not assumed the customer's motive for asking, I could have learned the real intent behind the question, responded with the better price I was willing to give all along, and won their business.

On one hand there is an unprecedented rate of innovation happening in most sectors now, and on the other hand all sales roles have effectively become telesales roles, thanks to the COVID-19 pandemic. Several sectors have been impacted negatively while several sectors have experienced a positive impact. This can be an incredibly complex set of variables in your territory and in your role, but you need to develop skills to navigate these complex variables and the decision-making should be based on the minimum number of variables you can. The twelve lenses I laid out earlier in the chapter can be incredibly useful when it comes to simplifying the decision process on where to focus and to identify the best path to win. Try it a handful of times, and you'll see a win plan start playing in your mind in a very short amount of time.

It's my hope that you glean from both my successes and my failures to lose less, win more, and have a lot more fun in your career. The five components of your win plan, along with the twelve parameters, will help you approach every deal with greater odds to win. From there, handle objections with care, tie up loose ends, and remember that no matter what happens, there will always be another deal coming along. Do all that, and you'll become the most successful salesperson on your team.

BE WILLING TO ASK FOR MORE

In sales just as in life, there is only time and energy. These are the only two resources you have to win deals, so you'd better be sensible about using them. In the context of your win plan, that means you must be willing to risk losing a small deal if there is potential for significant revenue from the account. I've taught salespeople this lesson wherever I've led. Most sales teams I've trained are hungry and willing to put in the work, but too often they find themselves with small deals, all but giving the product away. There is a solution. Go big and risk going home.

Usually, it's worth the risk. Here's why. I always tell new salespeople that $25,000 never solved anybody's problem. You have to go after $300,000 to $500,000 deals. When I started with Data Republic, the sales team in Singapore was working on a $25,000 deal in Indonesia. They planned to trial a proof of concept in hopes of earning the business. The customer would not give us more than $25,000, one team member told me. When I ran the numbers, I realized that only $25,000 for the service we provided would send our company into a loss.

When I joined the sales team on the next call with the customer to discuss the proof of concept, I told them we had to pull back. The relationship had big potential, I said, but $25,000 would be a loss for Data Republic, and we didn't want to do that.

"If you are serious, we will put all our resources into our relationship," I said on the call. "That means a full-fledged, three-year contract."

"Send us a proposal, please," the customer's final decision maker replied.

We came back with a three-year $900,000 proposal. They told us they did not have the budget for that, but they could afford $350,000 for a lower level of service that still provided significant value. The sales team was stunned; they never thought a $25,000 deal could turn into $350,000. In fact, they would have been happy with $25,000. Why risk losing even that? They needed to learn that both parties need the relationship for it to be worth it. As you draft your win plan, never be afraid to offer your biggest, best solution for a price that is advantageous to you, assuming you understand the customer's needs, of course. Everything you've learned about magical selling in earlier chapters is a prerequisite to winning. Follow this tip—be willing to tell the customer no to a small deal—and you will win big more often than not.

In that particular deal, we ended up earning more than ten times what we asked for originally. The entire experience proved that people don't have a shortage of money; they have a shortage of trade-offs. I will never pay $500 for a burger. That doesn't mean I don't have $500. I'll pay $500 to fly somewhere. But a burger, I'll pay no more than maybe $20.

That had better be a good burger.

CHAPTER 6

Micro-Lessons to Grow Your Sales Career

Chances are you have heard the joke about the oldest profession in the world. I won't reiterate the punch line, but I will challenge it. Illicit services do, in fact, require selling before they are provided, which makes sales the world's oldest profession.

Years ago, I attended a leadership training seminar. On day one, the trainer asked each table to pick up a slip of paper from a bowl and discuss the topic written on it. Our human resources director sat next to me with two people from marketing. When it was my turn, I picked up a chit with this question: *If there was one thing you would like to teach anyone in any role, what would it be?*

"Sales," I said.

"Really?" My HR director looked intrigued. "Interesting. I had never heard anyone say that before", agreeing with me while making that comment.

"Think about it. If you are interviewing someone to hire, you are selling the position, the company, the team. All great things," I said. "If you are interviewing, you are selling your skill, your knowledge, and your personality. If you want an employer to buy, you must be able to sell. Then, once you get hired and you want to try for a promotion or negotiate a pay hike, what must you do? Sell."

"I had never considered that before." The HR director nodded. "Success in any position is not a result of subject matter expertise alone, you are saying."

"Correct. You need to know how to sell well to get the job in the first place. It's not politics; it's sales. Whether you realize it or not, selling is inherent to every career. If you are an entrepreneur, you are selling your idea to an investor. The best salespeople ink the most funding. Any accountant, marketer, or human resources specialist with selling skills goes further, faster in their career than colleagues with only job-related knowledge."

Let's say you are a research associate and need funding. How does MAGIC work? There is always an equivalent of M—Money, Authority, and Need—to approve and support you. Work on A, Alignment, of the outcomes as a two-way street. G—Grow your influence with additional ideas, knowledge, or partnerships, followed by I—inspect your progress—and C—Curiosity or Context or Care.

If you want a promotion in your job, you can apply a MAGIC thinking approach to engineer your career path. Where do you want to apply this?

To me, there is no position more rewarding than sales. I love every minute of it, understanding and trying to solve big problems and getting paid. The better I get, the more I'm rewarded. No other profession connects outcomes to compensation this way. Salespeople who make their quota typically earn 50 percent or higher remuneration than employees with similar education and experience who work in the back office. As a result of a job well done, revenue increases, and you earn your own share. Over time, you can move from individual contributor to other operations with higher pay such as people management or profit and loss (P&L) responsibility. Perhaps you'll soon reach a position where you're training your own team—or training sales managers to train their teams. Opportunities abound. Selling is a life skill, but for some reason there are hardly any universities specializing in sales skills.

Whereas previous guidance in this book is all about the sales process, this chapter is about the profession itself. Sales has much to offer if you are up to the challenge to receive it. Sales mastery also opens the door for entrepreneurship. What do you think entrepreneurs do? The successful ones are good salespeople in addition to any function they are trained in or have worked in. They have a natural selling process, and that's how they position themselves better.

Sales requires skill, dedication, teamwork, and situational flexibility. And there is no second place prize. You come in first or you lose everything. Sales can be cruel, which means you must do everything you can to gain an advantage. The best product may not get sold, but the best perceived product will get sold. You as a salesperson are a product, in a way, irrespective of your profession, so I will teach you how to make yourself the best perceived product in the pages ahead.

ADVICE TO MAXIMIZE YOUR EARNING POTENTIAL AND JOB SATISFACTION

Here are advanced tips I share with new and veteran salespeople to help them take control of their careers. While this advice is centered around sales professionals, anyone who is selling anything (even an idea), asking for a raise, looking for a job, or seeking a bigger budget can leverage this advice.

BOUNDARIES ARE MEANT TO BE PUSHED, EVERYDAY

The thing about selling is that it doesn't measure whether sellers consider themselves right or wrong. What matters is whether they are effective or not. Whether they closed a deal or not. So what if you feel you did the right thing (according to you, of course) and your competition won? The right thing is also forever evolving in real time because no two deals are identical in B2B sales. What looks similar between the previous customer and the next one is what gets you to the table. What gets you the deal, however, is how you identify and manage the differences. Never be the same as you were yesterday. It means you are asleep at the wheel, and it's up to competition to let you win. Make that mistake, and your company and customer will segment or label you, which means you're limited by others' perceptions. The same goes for your role. If you've been a key account manager, you only get key account manager roles. If you only do small and medium businesses' deals, you get more but never enterprise accounts. So push your boundaries all the time. Work cross-functionally with others in your company to do work you might like to do someday.

What is right for one customer may be wrong in another deal. The right thing is always contextual, no matter how many similarities you see between opportunities. It makes you practical. It helps you take rejection better. Don't forget that the revenue earning function always needs to be the sharp end of the stick. If you are doing well, it automatically means today you are doing better than yesterday. Why is that critical? Because you need to play better than your competitors and they are getting better all the time and new competition is forming all the time.

One way to ensure this is to talk to at least one new customer every day (or someone from an existing customer's new department) and apply a little bit of MAGIC on that person. This is like exercising. You won't gain muscle by working out for six hours one day. But if you do it daily for an hour and, one fine day you'll realize you are in measurably better shape. When you speak to a new person, adopt a "does it make sense to sell to that person or not" mindset rather than "I must somehow sell to this person." If you don't, you may end up with happy ears but a waste of time.

GROW YOUR INFLUENCE AND PLOT THE GROWTH (THE G IN MAGIC)

This tip is the corollary to the first. You are what you do; you are not your job title. The onus of proving yourself is on you. If you are not happy with the way your company, customers, or even competitors perceive you, actively change that. Push boundaries on what your sell, who you sell to, how many you sell, who you partner with, etc.

Most don't take initiative to push boundaries all the time because they get comfortable and afraid to fail and get hurt if they are not able to get things they want. I suggest that everyone should do sales in their career for at least a few years. It really improves your perception of life and situations in general and gives you the ability to adapt to any situation.

STOP RELIVING THE GLORY DAYS

Every once in a while you win a deal that is worth bragging about. I did this many times, and many others around me did too. The issue is when you are stuck in the glory days. One of my favorite ways to bring myself or anyone on my team who did a fantastic job in a deal back was to say, "All right, you've earned yourself thirty seconds of bragging rights. Your time starts now," and I'd look at my watch. It works in a sense that it brings you back from living in the past and grounds you in what you need to do now. In a salesperson's career, every quarter starts from zero, irrespective of what happened in your previous quarter. So learn to start from a "zero" mindset. Use all your past experience to get to the table and improve your productivity in various aspects of sales cycles. You try something. It works well. You try it again. It doesn't work. Pay attention. In the sales context, most people keep trying. Maybe they work in a different context.

In any difficult situation, the typical refrain goes a little like this: "We used to do it this way."

If people constantly talk about what used to work, that means they aren't making an effort to adapt. They're not making any effort to understand what to change. Make the mistake of doing what used to work even when it doesn't, and your career will taper off long before you realize your potential. Every time I see this behavior, I know what happens next. I've predicted at least three managers being moved somewhere else because they refused to understand what shifted. Maybe the reassignments turned out well for them, but they were misfits.

If you happen to say, "We used to do this in a previous company," that's your cue to stop reliving the glory days and start creating a new future. If you hear someone else say it, this phrase can also operate as a tell that the person may soon be quitting their job.

PAY ATTENTION TO WHY CUSTOMERS BUY (AND THE REAL REASON DECISIONS ARE MADE)

In each and every role, I have always tried to understand the solutions people want to buy. It's rare that I see someone buying stuff for no other reason besides solving a problem. Still, there is a gap between any product you're selling and what the customer wants to buy. Remember, you are selling a drill bit, but the customer is buying a hole. Ongoing product education will enable you to make that connection quickly and, from your company's perspective, profitably.

One such example was a city mayor in Southeast Asia who wanted to put an end to corruption and fraud across several agencies. We were selling an identity management and content management solution. The mayor assembled the heads of fourteen difference departments along with technical employees and managers. One challenge was the mayor understood a little English while the others spoke none. A local partner was a translator.

"This works," the mayor said after our slide presentation. I had spent 80 percent of my time explaining how we could solve their problem and the rest on features and functions. We got easy buy-in from all department heads because they simply wanted the problem solved.

People will tell you that selling into different cultures, languages, and behaviors requires different approaches. I disagree. Simply be aware of every customer's needs. The bigger the problem to solve, the higher priority you are. Never sell to a nonessential need. Think about TINA—there is no alternative. Does the customer have no choice but to solve their problem? If not, don't waste time. Step out of all unqualified accounts. For qualified customers, align your product's features and functions with the solution they want to buy most, knowing that the same product can solve multiple problems. It's your job to find out what the customer wants your drill bit for, then to sell that to them.

BRING THE BOSS TO WORK

Once my boss once asked me to pick up *his* boss. The gentleman, whom I respected greatly, insisted on talking to customers to get feedback on how the sales team was doing. I decided to take him to a customer whose deal had stalled. When we got in the car the big boss asked me which customer I was taking him to and jokingly added how much I "paid" that customer to give a good reference when the boss asked. I told him that he could check with any of my customers any time and I wouldn't have to waste my time accompanying him. But I was stuck in an account and couldn't seem to find a way out. "I'm taking you to this account so you can help move the deal forward because I can't," I said. "I want to see how you do it differently so I can make my number and learn a way to deal with such deals." We made some progress in that meeting and eventually got revenue out of that account. This is just one example of leveraging the boss. For anyone in sales, I would recommend that you have a plan on leveraging your manager, irrespective of whether you are an individual contributor to the team or a manager. There are several benefits to doing this proactively:

1. When you are stuck in a deal or in your thinking, they can get you out objectively as they have likely experienced similar issues before. This saves time.
2. Economic buyer coverage. Executive relationships will help you check if the deal is on course. They may not be so accessible to frontline sales teams or even first line managers.
3. Internal positioning. Some bosses are operational by nature; they are more useful to leverage for internal stakeholder management than external for a variety of reasons. The point is how to advance to next steps. What works best is when your ask is specific. The more specific the ask, the easier it is to leverage. I have noticed that many sales reps and managers confuse this with sharing their problems. That may work sometimes, but that's not what I'm referring to. State the problem, and as far as possible do the thinking first on how to solve it. Then ask for suggestions on which options are best or for approval of what you are doing. This little act done well regularly will set your career up for success when the next opportunity shows up. It's a simple skill, not politics as most people mistake it when they don't look closely.

Keeping bosses in the loop on how you are thinking of addressing an opportunity or a territory or a certain problem always helps them look good in front of their management. They will appear to be in more control of the team and business. Sure, you will have to be clear on how much information to share because not two managers are the same, but doing that to an optimum level will make their life easier. For others who might not do this as well as you do, it may look like politics. When someone says it's internal politics, it is more often than not lack of skill on the part of the person complaining. I have practiced with more than a dozen managers to just speak about work and avoided any other aspects but still ranked high in their ratings (highest in most cases). So before you complain that your boss is political, ask yourself what part of selling you need to improve in.

You want career growth. When your manager is more involved with you, chances are you will succeed faster and it will lead to more doors opening in your career. This skill is also called "managing the boss." Just like you manage your partners, team members, customers, treat your boss the same way.

BECOME A SWISS ARMY KNIFE

A Swiss-army-knife salesperson is useful in every selling situation. If you're only getting one sort of experience over the course of several years, you are only specialized in that area. This limits your growth potential. But if you are a Swiss army knife, you're flexible. So think of ways you can sell different products or partner with other people to strengthen your ability to solve issues, bring results, and close more deals. For example, if your company offers more than a single product, learn to sell them all. And if there is a new product, take it to the market before your peers. Be the first to take any new product launch to the market to check for potential interest. Do the same if your organization acquires or merges with another company. This was my favorite approach to understand customers better, differentiate from competition ,and build bridges within the company while I was at Oracle. Don't underestimate the skill you get and the roles that might open up for you because of that. In some ways this future-proofs you from being fired for any number of weird reasons, such as the pandemic. I've seen companies get rid of worthy candidates by just looking at the spreadsheets. Give your organization a strong reason not to include you in such scenarios.

To do that, ask yourself, what does it take to be successful? How do I measure success? The same qualification you use with a customer is how you can decide to add skills to your CV. You are the CEO of your life anyway; you need to ask these questions to find out how you can measure success. If your measure of success is to double your pay, look for places where that desire can be met. And if you want to do that in a role you haven't done, you need a system to get there. You have to plot from where you are to where you want to be. Sometimes it could take two or three steps. Sometimes it could be more. On rare occasions, it could be just one.

There are more people coming into the job market all the time. You won't be the only salesperson looking for the best job, but if you are a Swiss army knife, you have an advantage. You know more because you've done more. It's a talent stack.

Along the same lines, if you want to move roles to increase your pay, go ahead, and don't underestimate the need to continue sharpening your Swiss army knife in the new job. Your Swiss army knife background is what gets you the job in the first place. On your CV, include examples of 100 percent achievement across small deals, large deals, strategic deals, competitive win back deals, and new technology deals.

FOCUS ON MAKING YOUR SOLUTION EASY FOR CUSTOMERS TO DEPLOY

Reducing and eliminating friction in your dealings is the secret sauce to win. If customers think you are easy to deal with, your solution is easy to deploy or learn and easy to get results, you will win the deal most of the time. No one has the time or patience to run the elaborate processes that were common even a year ago. The pandemic helped propel the efficiency and simplicity game to new heights. Give yourself a better rap so they know you are useful for their business.

Then when the customer buys, stick around to check monthly progress. You don't disappear with your commissions. You're a partner they can trust. That goes a long way toward building partnerships with the customer's team, which results in cross-sells, upsells, referrals, or all three. This is even more relevant in pandemic times. Customers will initially buy from existing vendors in smaller chunks, get quick results, and repeat. This will continue until customers get used to buying remotely even for large transactions. We are already seeing large RFPs issued and briefed to dozens of bidders at the same time. Once the solution is submitted, the technical and business evaluation sessions are scheduled (no pricing is required at this stage), there are steps like demo or proof of concepts to go through before a final list of vendors to submit their pricing. Selling careers changed forever in less than a year. One approach we tried is to adapt our pricing methodology to the customer's way of measuring success. Unless your internal processes are simplified, it's hard to simplify your customer's experience with you.

SPEND YOUR TIME WISELY

In other words, work smart, not hard. It's a simple difference. Hardworking people focus on doing more of the same—more calls, more presentations, more of everything they are doing now. Smart workers focus on making effective decisions and end up becoming more productive. One of the most important decisions any salesperson must make is how to spend their time. Specifically, how much time you spend in three broad areas: discussions with customers and partners, operational work, and system updates. Operational work includes responding to RFPs, looking for and assembling content, etc. Did you notice that the Pandemic seems to have increased operational workloads on everybody around you? Why do you think that is? It's probably because you are still fighting with your sword when it is now a gun battle out there. Not a smart move. It means that you need tools (guns) to automate your own RFP responses. From taking meeting notes to automating RFP responses to sales enablement there are tools available now. If you are not using them already, you risk falling behind. Free up your customer facing time and learning time. Try to automate everything else with these new tools.

If you need "guns" in your specific case or for your team or your organisation, reach out to me for a chat.

In sales and in life, rarely do events go according to plan, so set your agenda in a way that's directionally accurate. For example, a lot of companies are keen to get higher margins at a transaction level but don't give much thought to their "land and expand" strategy. That gives the customer the impression that you are hard to deal with, which makes them play defense. It delays deals and increases your risk of losing. There is a high-profile bank that wanted to renew their existing vendor but found our product to be better. The pricing was below our usual selling price, but we managed to justify that lower price in the interest of time and the opportunity to expand. So the plan is directionally accurate, and with resources allocated for customer success with monthly checks. Once you do that, you easily find out what are the adjacent problems the customer has and that you can solve them. That's the expanded part that needs monthly reviews.

ARE YOU BUILT TO LAST?

Smart people focus on making decisions such as writing their win plan. Everyone else focuses on working hard, which is doing more of the same. To be fair many managers out there push their teams to make more calls, build more pipeline, burn more hours. Reps may end up cutting corners and managing upward instead of getting smart. Which one are you? If you don't believe it, just pay attention to all the successful people around you.

Many sales managers try to motivate the team with, "Work hard and play harder," whatever that means. Why should everything be hard? Hard deflates people. I was a part of that environment for many years. I followed the suggestion anyway. Eventually, I realized that working hard made me a machine. Did you ever get a feeling that if you work hard they will always give you more work or put more pressure on you? I know I did. In one of the roles early in my career, I outperformed everyone on the team and was expecting a promotion to a people manager position on account of my performance. It didn't happen. Somehow it happened when my results tapered off a couple of quarters in a row. Later I joked about that. When I was doing well with numbers they gave me more work, but when I delivered less they decided to make me a manager because I'm no good! A cog in the wheel can be replaced any time. Hard work means more of the same. Smart work means spending more time on better decision-making. I have always tried to push boundaries so that the deals always are interesting for everyone involved and winning is always fun and energizing. The secret is that if you keep pushing boundaries on every deal you do (e.g., negotiation or partnerships or sweep the table) it means you are built to adapt. Whoever is built to adapt is built to last. Those who simply oil the machinery are just one disruption away from losing their job.

FIX YOUR COMPANY OR FIND A NEW ONE

If you ever find yourself talking about how great life was at your previous company, that's a sign for you to change jobs. If you don't find yourself pushing boundaries, if you're too accustomed to your habits, either fix your company (and yourself) or move on. Salespeople are sometimes comfortable in their space, but the ability to adapt to multiple situations, companies, and cultures gives you wisdom to find a better opportunity.

The catalyst may not even be you. Perhaps management loses their ability to understand the ground situation. Yes, that happens a lot. If you noticed your bosses only "preach" decisions instead of staying close to ground and making the moves, then you definitely need to evaluate what you are doing there. All my moves happened when I realized management was not willing to listen to field salespeople's feedback. We used to call them CEOs, Chief Excel Officers, staring at spreadsheets all day. They didn't understand why some salespeople struggled, and they did not look to help. Voices of the ground team went unheard.

You, too, may notice senior leaders you look up to becoming theoretical over time. You'll notice their decisions show they are losing touch with customers. That's the first sign to be alarmed. How long does your company management need to, say, adjust a sales rep's territory from the time you realize it and ask for it? Some big companies take months!! I'm not saying jump ship, but it is time to look for a raft. If your management is disconnected to the ground situation, look for a job outside your company.

As long as you stick around, it's your responsibility to push boundaries. Ask questions your out-of-touch leaders won't. What's the territory like? How has everyone performed in the past? What does it take to reach a new level of success, both from people within the company as well as external partners? Management may not listen, but you will be sharpened and focused if and when you get hired somewhere you are appreciated.

ALIGN YOUR STRENGTHS TO THE RIGHT JOB

I've had to ask more than a few salespeople to quit. When it comes to that, obviously something is not working well. I may want to earn millions playing football, but what if I cannot strike to save my life even if there is no goalkeeper? I'm a misfit for the pitch. Perhaps I have a natural talent for haircutting. I would be better off trying to earn a fortune styling celebrities. In other words, I should look at my strengths and find the best opportunity for me to use them.

Few people are good at everything. While you do want to build your skills and knowledge base so as to become a Swiss army knife, it is better to do that in a role where you can excel than in one where you struggle because you won't make money, and you'll look bad.

If an opportunity plays to your strengths, you will win. It may take a year, maybe two, but you will get there because it will not feel like work. You may be the math wizard who thinks calculus and physics are easy. Most other students will disagree! The alternative is a job where your pay and confidence are low and stay that way.

ARE JOB INTERVIEW SKILLS A ND SELLING SKILLS ONE AND THE SAME?

Throughout my career I advised many, many people on job interviews. One of the main reasons why competent people don't get through interviews is because they don't approach it as selling. The difference is you are selling your skill and will for a certain job/pay.

When it's time to change jobs, reject the fear of missing out. Too many salespeople have bought into vague advice like looking confident; being bold, assertive, polite, and energetic; being yourself (I've never understood how you can *be* someone else!), hoping for the best, mirroring your interviewer's body language, and whatnot. You want results. The only thing that matters is preparation. Those adjectives are really a byproduct of your preparation. But how do you prepare? If you practice MAGIC in your daily activities, I'd say you can be ready on short notice. Still, I'd like to give you three things that can help you prepare for your big interview.

1. Where is the money? (Accounts, verticals, territories, countries, etc)
2. What is your network in that patch? (Accounts, decision makers, partners, etc)
3. How do you plan to get that revenue? (ideas, market intel, help required, recruiting etc)

Think about how to apply each of these three to your situation. Either you:

1. Know these answers for the role (in which case you lead the conversation) or
2. You don't know (in which case you find the information during the interview while sharing how you do this or have built this in your current role/territory) or
3. You are somewhere in between (in which case you should lead where you know and gather information where you don't know).

If you have to prepare a business plan, just say what you would do in these three areas within a certain period to produce the desired outcome.

Now, how do you ask intelligent questions during an interview? Remember, interviewing is nothing but selling yourself. So the answer is MAGIC, of course. Ask how they measure success (revenue and non-revenue). What is it now? What do they want it to be? Ask yourself the same questions. What makes you interested in the new job? Bigger territory? Bigger pay? New team? Different challenges? Dream company? The opportunity to play to your strengths? Make a list and state it upfront, to the point.

It doesn't matter if you're going into sales management, account management, or a sales rep job. If your intention is to understand what's going on, your prospective employer will appreciate that and will want to hire you, even if you're applying for a role you've never worked before. I have seen account managers get hired to run the entire Southeast Asia market. I've also advised and seen individual contributors get regional roles directly. Including a large territory like APAC in one move. It's not easy to get into those roles, but smart preparation helps you work at a higher level.

Stay on point and execute on point. See the MAGICAL results for yourself and reach out to me if you need help using this simple system.

NEGOTIATE YOUR SALARY TO WIN

Over the years I learned a few tricks of the trade, but clearly a vast majority of people are not confident to negotiate their pay higher. Even the ones who negotiate very well with customers and partners are sitting ducks when they want to negotiate for themselves. Several people—not from my team but within the same company—have approached me on how to negotiate their salaries. Why do you think that is? The primary reason in my observation is they treat it differently than a customer negotiation. Most assume that if they are nice or work hard, their managers will get them a pay hike. This may be true in some cases, but it is not a rule. In any job role - not just in Sales - the reality is that you have to do regular checks in the market about what kind of roles are available for the skills that you have and for the skills that you can easily upgrade to. Not doing that means you limit your ability to say "No" to any unreasonable work and ability to ask for a raise. And don't blame the company or your manager for making you feel like a slave. If you feel like you are slaving it away, you really need to work on your selling skills in addition to whatever is your functional skill. That is even more important in the current disruptive environment.

As with customers, salary negotiations are also individual-specific, but you can have a system that helps you on the way. You need to recognize that you are selling yourself—your skill, energy and time—and you are the product here. Sales is a rejection business. You hear more nos than yesses. Therefore you should be able to use the MAGIC process.

First, understand your boss/company. What is the general policy of pay hikes and promotions? Know the rules. How do they get measured? You need to know under what circumstances there will be exceptional pay hikes or promotions. Many times it's a matter of asking what the rules are. (This is the M in MAGIC.)

Make sure your manager knows that you understand them better than others. This is Alignment. There are simple ways of helping your manager stay up to date with your work. Make it predictable for them. Maintain good CRM update hygiene. Describe it in a way that allows them to use those examples upstairs, with their own bosses. Set lower expectations, but make it predictable by giving the range of your request and providing reasons for that range. Performance is the key as usual. But there are instances where other issues may require your manager to spend valuable political capital to defend you. Don't let that happen if you can help it. Your manager's political capital must be used primarily to increase your pay.

The biggest stumbling block I've observed in salary negotiations is this: the wrong expectations. Use the MAGIC framework every day to get results and help your case. That doesn't mean performance always leads to a promotion or raise. Sometimes it's simply not in the process to offer a raise. Unlike deal wins or losses most reps think that a pay hike rejection is a personal rejection. It isn't. The fear of rejection doesn't allow most people to negotiate higher. If you don't let your ego sabotage your progress, which usually results in you operating either from a place of fear or of greed, it's an easy skill to acquire.

Simply use MAGIC to find out what it takes to get a pay hike. What's the metric to be considered for raises? Set the expectation that if you achieve it, you are expecting a pay hike. Don't beat around the bush. Don't laugh or smile after you ask. Just listen. If you have to follow up, ask pointed questions. As with a PO, wait until you get within striking distance of your goal before you remind your manager. A simple way to say it is, "I've held up my end. What about yours?" Same with a promotion. Put that in writing politely for inclusion in your performance assessment.

HIRE FOR WILL, NOT SKILL

I'm saying don't compromise on the "will" part. If you have not already, at some point in your career, you will be tasked with hiring your own salespeople. And sooner than you think. I'm merely saying skill around a certain solution can be acquired. If you get a 100 percent fit, it's good and good, just hire the candidate. But that's rare. You should know now that even the best candidates will only bring part of the skills you expect. To be perfectly functioning on the skill part is fine because people do learn on the job. The part that you need to be more particular about is the will. Ensure there's no compromise there. In fact if I think it may be a good idea to actively look for candidates who are one degree away from what you want them to sell instead of having an identical background with a different company. The reason is that, within months, once they learn the new stuff they will be far more effective in managing customers. Anyone with the will start delivering in less than six months in my experience. That's why I always consider professionals with a one degree of difference for sales roles. They also won't get bored selling an identical product of a different brand. I have hired some technical people into sales, individuals who never had a chance to do formal sales roles before. This isn't rocket science. The fact that they understand the product and the industry better than nontechnical salespeople gives them a distinct advantage. All they needed was awareness of necessary skills and the chance to build them. Joseph is such an example. I gave him the opportunity to become sales director, running a team of twelve people. Sales director is not an easy position because the work also involves managing team members, managing forecasts, managing reports, and troubleshooting anything that goes wrong.

Still, take a realistic view of every candidate's abilities and potential. I have received my share of headhunter calls only to realize they didn't want to move forward after the first conversation. I was intrigued as to why someone would reject a hardworking guy who can generate results. What shocked me was who they hired after rejecting me—someone who didn't last a year. That's when I realized those hiring managers didn't have the skill or the expertise to identify real talent. Sometimes it looked like they didn't want intelligent people who might expose the boss's lack of depth. Just take a look at any job description. They want team players, relationship builders, overachievers, and high-energy professionals with multiple decades of relevant experience, discipline. Once I saw a job description that read, "Not achieving sales quota is like blasphemy." That company went out of business a few years after I saw that job description.

Hiring managers who ask for Superman or Wonder Woman end up hiring wimps in many cases. I've told as much to headhunters and hiring managers who've reached out to me even recently: "Your job description looks like you want a superhuman salesperson, so I'm not interested in wasting your time or mine."

Do not make their mistake. Hire for will, and be flexible on skill. If a salesperson has no will to learn or put in their best effort, it doesn't matter how much skill they have. They must show the ability and the desire to develop their skills. That's why ramp up time is important. If they can pick up knowledge about the drill bits and the holes in that time, the results will be productive. Technical people who enter sales with a will to win (and a willingness to handle loss with grace) become incredible employees who end up teaching you just as much as they learn from you. It's a fantastic experience.

LET PEOPLE PROVE THEMSELVES

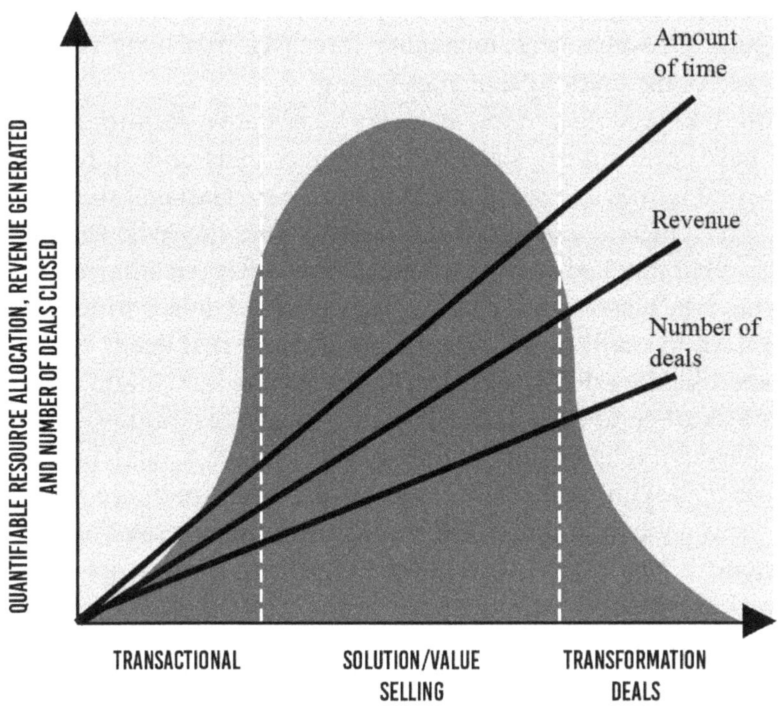

Figure 12.1 The Three Types of Deals.

A rep who predominantly sells transactionally can technically be promoted to handle the more difficult territory or accounts. Look for these opportunities and allow your reps to grow. If you are a rep, consider volunteering into other territories. In your CV, mention how you close business in your role now. There is nothing wrong with being in one business or the other. It's a matter of fit, and if that business is playing to your strengths and aspirations, stay.

These curves are also different for different companies based on their solutions. That's why there are differences between and among hardware sales cycles, cloud sales cycles, application sales cycles, and so on. A rep who is selling transformational deals in one business may be considered a transactional sales rep in another company.

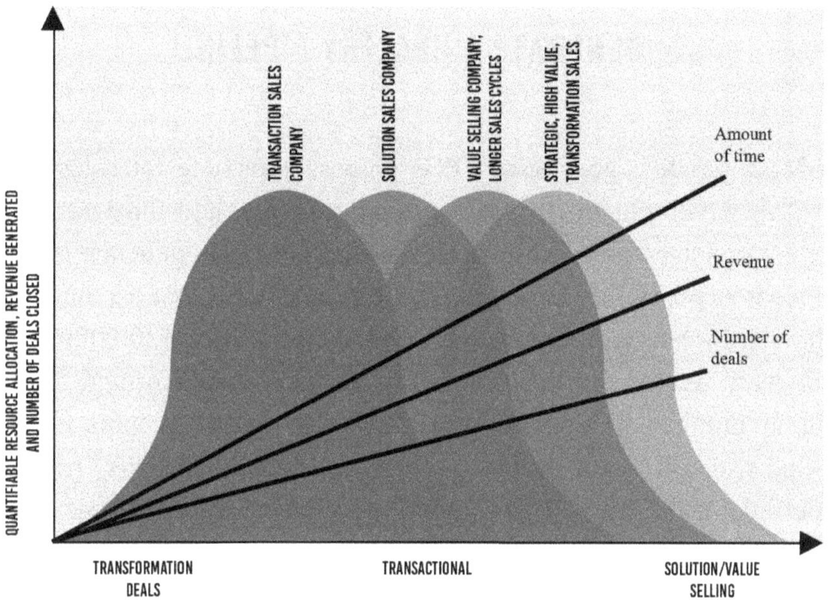

Figure 12.2 The Three Types of Deals (with overlap).

As you can see, there is an overlap between the sales processes. The difference is on the extreme ends. In a sales career, it's important to get these different perspectives and experiences. Experience doesn't mean you should spend years selling each product. Partnering or paying attention to how customers buy, how colleagues upsell, and how partners work with their accounts. If you have this perspective, it's easier to predict if a particular job is good for your career or not. And how successful you will be. Who to hire, how to coach, how to pitch, etc. all become easier.

The above diagram shows how different types of companies operate in their own spaces and how you can position yourself the way you want. Just sit back and look at where your new employees are coming from and when they quit—and which companies do they typically join. You might see a trend there. This mechanism is purely to allow yourself to create opportunities for your team or yourself to further your career. That's how you can play to your and each of your team member's strengths.

GENERATE A HEALTHY PIPELINE

Most pipeline generation (PG) processes include telesales teams, business development people, marketing events, and third-party contractors and agencies whose shared assignment is to generate as many hot leads as possible. Many organizations have KPIs that encourage bad behavior. I've seen BDMs being asked to add at least $200,000 in new pipe every week. It's a massive spray and pray endeavor and reps just cook up numbers to bloat the pipeline. It also doesn't account for some ground realities. For example, government departments may not allow hyperlinks to be opened from emails, so all email campaigns are not productive. In several countries if an email is in English it's pretty much guaranteed that they won't read it. All these activities are unproductive, but these are the rules management puts on reps to generate pipe. How will it turn out? You probably hate PG because your tasks involve hundreds of calls, emails, and direct messages. Often, these efforts seem unproductive because results are unpredictable. When sales directors realize the last quarter was awful, they demand a 4x or 5x pipeline, which you know is inefficient. They probably do, too, but their boss wants the sales activity reports to make them look good. It's all optics so the bosses can keep their jobs at the expense of your time.

To succeed at PG, we need a simple mindset. That way, I've found, is the three-bucket method.

- Account-based pipeline generation
- Partner-based pipeline generation
- Generic pipeline generation

Let's cover each in depth.

ACCOUNT-BASED PIPELINE GENERATION

Account-based PG is lead generation that targets specific, high-value prospects. These are usually nominated by the Sales team. It can work when you have an existing account with significant revenue and decent potential for more. These are generally existing customers. The sales team generally knows the people, process and technology and therefore is able to target specific outcomes in a predictable manner. The investment may end up being high but when directed accurately and executed well, the result can be very rewarding. You not only get to upsell and cross sell but strengthen your customer relationship to ward off competition.

All key accounts should have a plan for such activities for PG monthly or quarterly. Don't hesitate to suggest this to your management if this forms your main patch.

PARTNER-BASED PIPELINE GENERATION

Many salespeople do partner-based pipeline generation instinctively. Recall from the partnerships chapter how your champions will refer you to their customers or vendors. That's partner-based PG. It's a slow process. Once you partner with a company or individual, it takes months to learn each other's strengths, skills, and solutions.

There is a large SI company that secures hundred-million-dollar government contracts. It's a company you want on your side because, despite all the complexities associated with a complex billion-dollar organization, the advantages are worth it. To partner with them is to receive several of those projects almost effortlessly. We had to start somewhere..

We brought one hundred presales people, account managers, and technical personnel to our meeting with the SI firm. We did a morning ice-breaker session with the sales teams and an afternoon session with everyone else. This was six hours total for them and for us that took 5 months of extensive work to organize. It was worth it because this orientation gave the SI company a look into our capabilities and our team a preview of the type of work we might win.

Not only did the SI firm refer their own projects to us, but they were a source of referrals that amounted to a significant partner-based pipeline.

This example gives you an idea of what success looks like. You could do that on a weekly basis with a different partner. Think of casual discussions among a few people over a beer. You don't need hundreds of people crowding an auditorium. Whatever works for you.

Again, this is a process. It's not like you make one phone call and another company sends you leads. It involves time investment, but not nearly as much as account-based pipelines. That might mean, for example, your partner invites you to internal meetings where feature updates are discussed. It helps you understand your partner better, which ultimately produces results.

GENERIC PIPELINE GENERATION

Generic pipeline generation is casting a net to see what happens. Think webinars or sponsored events to which you can prospect. We held free breakfasts and invited many customers to come learn about a relevant technology topic while they ate.

I don't expect these three pipeline generation buckets are new to you. You have heard of them, but may not have done all three yourself. However, it is the industry standard to fill your schedule with all three. The key is to organize the ones relevant to your patch. Building an ecosystem has its benefits.

If you are early in your career or are a new hire, you may not have any partners yet. Still invest half your time in seeking and building partnerships. If that is the case for you, refer back to the partnership chapter with an eye for advice about forming referral-generating partnerships. I also recommend you adopt the hunter spirit. For example, you could attend a generic pipeline event such as a trade show your company sponsors, then pursue each attendee as account-based pipeline generation.

Whatever your schedule looks like, manage your boss. Share your plan, and state your expectations. If you're a sales manager, give your team the template and allow them to think through execution themselves. Allow their ideas to be heard. That's how salespeople get engaged and contribute more. Managers would do well to help each team member realize their potential.

THOUGHTS ON SELLING DURING COVID-19

In a nutshell, COVID accelerated every trend that was already taking shape. There was a trend of working from home, and in the technology industry everyone stopped going to office for a year or more. If your trend was negative, even that got accelerated big time. Jobs that were not pushing boundaries simply disappeared overnight.

I'm reminded of different points in my career that coincided with some kind of disruption. As mentioned in chapter 1, I had a preplacement offer in the summer of 1997, but there was the Asian Financial Crisis. That led to several companies canceling out campus interviews, and the company that offered me a job was bought over. My job options looked bleak during my final semester. It led me to take up a temp role in photocopier sales for my remaining few months in college, which was my first exposure to sales as a career.

Everyone remembers the dot-com bust in 2001. I was selling a "cloud" solution for e-learning solutions at the time. Both cloud and e-learning were industries in their early stages that made it hard to close more than a handful of accounts. Those battle scars and the related recovery got me into Oracle, where I would spend fifteen years in a variety of sales roles in different countries and solutions in a variety of contexts.

Yet another big change happened in 2008 during the subprime lending crisis, and despite some fantastic results (>140 percent the previous year), I "lost" my job and got a new territory to build from scratch. I was shocked at the mistreatment but pressed on with delivering. Little did I realize that this brought me to the next orbit in my career for a full territory responsibility that would put my career in a different trajectory.

The "mini crisis" was that emergence of the cloud and some disruptive technologies. They were why I wanted moved out of a large company, a decision that led me to work in setting up and building new businesses at the cutting edge in various other territories.

And now we have COVID. Early in 2020 I was looking after APAC sales. It's a Startup, but the changes induced by COVID in the company's reorganization gave me the opportunity to become head of global sales.

While I was focusing on pushing boundaries in every role, none of the big turns in my career happened in an organic, planned way. It feels like "pushing boundaries" in your current role seems to automatically create opportunities for advancement when a crisis hits. (This is part of the G in MAGIC, in other words, embracing change as a way of growing your influence, knowledge, and experience in life.) The objective of this book is to help prepare all enterprise sales professionals to do things that will help set you up for success when a crisis arises. As they say, luck is nothing but opportunity meeting preparation.

Anyone doing well in a given circumstance but not pushing the boundaries of their roles constantly, tends to move sideways or effectively disappear when a crisis hits. This is blindingly clear to me.

The novel coronavirus, called COVID-19 by the World Health Organization, is the most documented event in human history. Everyone has an opinion on it. A perspective. A rant. I, too, have those, but I'd rather share with you my thoughts on and experiences with sales in our socially distant world. 2020 will go down as the most transformative year in enterprise selling careers ever—I think in a better way.

Let me explain with a parallel. Most of us drive cars. Years ago we had manual transmissions, and now it's mostly automatic transmissions. Not every driver was comfortable or good with manual transmission, and automatic transmissions allowed more people to drive by simplifying the work. In other words, it leveled the playing field. Similarly field sales is a more difficult skill than virtual selling because of various advantages of in-person selling. It also plays to some people's strengths, just like the manual transmission does. What I'm getting to here is, did you know that the insurance premiums effectively halved for automatic transmission versus for manual ones? A big part of that reduction was due to the fact that removing the stick and the clutch allowed drivers of all types to focus on traffic instead. The situational awareness almost doubled, thereby halving the probability of accidents. That led to a lower rate of claims and therefore lower premiums. Back then the discussions were around which car was better to drive, but no one paid attention to nearly double the situational awareness and half the premiums. I think the resulting elimination of the "clutches" and "gears" in the field sales process (i.e., time lost on road, between meetings, inability to meet customers in two locations) will improve the situational awareness of your deals, allowing you to double your or even triple your win rates. If you know how to use the new technologies to "double" your situational awareness to align to the new buying habits of customers, doubling or tripling of commissions is music to my ears!

While the news is all doom and gloom, there are definitive things that one can do to use this situation as a launchpad to get to the next orbit. The pandemic has impacted all industries but especially travel, transportation, entertainment, and hotels and resorts. Millions have been infected, and millions have died, but a host of others are OK. They work from home or work from their office but do not allow nonemployees into the building. Nobody wants to meet anyone from overseas in-person meetings anymore, much less with someone from across town. Unless in-person attendance is mandatory, video conferencing it is. So the question is, what must you do to meet your quotas and make your commissions in the new normal?

During the early 2020 COVID-19 outbreak, I was in Sydney, Australia. With my travel schedule brought to a complete halt, I had time to give my pipeline more attention than ever. Crisis is not the time to slow down everywhere. It's time to be aggressive in understanding how each of your customers' prospects are trying to adapt. COVID has closed the physical channels but opened up the digital channels. For many companies, that suddenly exposed how insufficient their digital channels were. In order to run digital channels effectively, the data economy is the key. The end game is to a) gain end-to-end visibility of supply chains, b) have full observability within an organization with the ability to intervene and course-correct operations, and c) also gain full transparency into the demand chain. It's a golden opportunity. This ties in to the first two Acts we discussed in this book. We sent out newsletters and webinar invites. It's easier to get someone's attention when they have nowhere to go but the kitchen. For deals that stalled because I misunderstood the customer, this has been the perfect time to fix it because they were not making any quick decisions either way.

There are three areas on which to focus during this time of disruption:

1. **How your market is evolving.** Focus on sectors that are growing. Yes, for instance, the technology sector is generating positive news. Travel, real estate, and so on, not so much. Customers are getting used to buying remotely even for large deals. Several customers want to accelerate purchases for their "priorities," and no one wants to spend the same amount of time they did pre-COVID. If you are talking to a new customer the quickest way is to go with partners who have strong, ongoing projects with them. If you are reasonably priced to keep things below the radar (meaning below the threshold of an RFP requirement, for example), partner well, you can sell a ton of stuff.
2. **How you need to evolve and approach AB testing on qualifications.** Pay attention to negotiation styles remotely. If you look at it, your selling time just increased. It's common for field

sales teams to be on the road for up to three hours daily. That's the perfect time for you to build your skills to sell remotely.

3. **Tips and Traps:** Simple discounts will be counterproductive soon, even if they appear to be required now. Find ways to ask how customers are now measuring their project outcomes and use that to fine-tune your pricing.
   a. Negotiations are evolving, so modify aspects to accommodate for virtual selling. Some negotiations worth hundreds of thousands of dollars are purely happening on email!
   b. Incremental deals are moving faster with existing vendors, so build bridges with those vendors as your partners..
   c. Sweeping the table is even more relevant. Of course you do that with partnerships, better network, intel sharing, and so on.

We had to abandon all agendas to close deals by a certain date while our customers were assessing the damage done to their businesses. One customer we were chasing for a $5 million deal announced a freeze. They would not spend more than $10,000 without approvals from two levels above. There was nothing we could do to change those limits at that time. Still, in most companies, deal progression continued. We just needed to allocate our resources until customers got the necessary approvals from management.

Since lockdown began, I found myself more efficient overnight. I was not on the road anymore. Instead of meeting four customers a day in person, I could hold ten to twenty virtual meetings. What would have been forty-five-minute meetings are now twenty-minute video calls. I've simply needed to be more articulate, concise, and relevant. It's not that I'm not the only one who can't meet customers. No one else can, so that's a level playing field. It cuts both ways, of course, so calling customers in the United States was not odd due to time zones. It didn't matter, and that played to my strengths. Sticking to strengths and focusing on matching them to the opportunity in front of us is the best we can do in a crisis. As for what to talk about, following the magical selling framework has not steered me wrong.

If you are in telesales, the rapid adoption of virtual business has been perfect for you. But what if you were in an impacted industry? If you sold to airlines, for example, you'd look for the individual responsible for ancillary revenue. Perhaps your solution could help the airline with a revenue-generating alternative to ticket sales. There is nothing wrong with taking advantage of the upsides of social distancing.

I also used the time to complete tasks I ignored because they were important yet not urgent. This included updating our CRM system. Initially partners had a lot more time on hand because they're also not meeting clients as often or as long. We checked in with them to see how they have been holding up. I also asked select customers if we could work together on a case study video about their benefits from our product. Of course, we made sure all customers heard from me, not just those who could provide testimonials.

COVID-19 presented a rare opportunity to grow your pipeline much bigger simply because you had more time to cover more opportunities. The world moves quickly. If you have not adapted to that change, I recommend you do. You can come out of this and any future pandemic not only with your health but with growth. So improve your efficiency. Reduce unproductive engagements. Let deals progress on their own. And spend more time with family.

I'll see you on the other side.

MAGICAL SELLING GIVES YOU AN UNFAIR ADVANTAGE—USE IT

In sales, we can't always control the cards we are dealt, but we can control the way we play the game. This is true regardless of the stage of your career, your solution, your company and your market, or your team. Some salespeople have a head start over others. Some have found themselves at the front of the pack. But even if you are coming into sales with no inherent advantage, magical selling will unlock opportunities you once thought were available only to the privileged.

As soon as you start making small adjustments, you'll notice a magical change. Of course, change doesn't mean anything until you're closing deals, but as you keep getting things done, you'll suddenly realize that you've always been a superstar. You were hiding, but now it's your time to shine. Your win ratio goes up significantly. Your ability to qualify leads improves, as does your ability to make winners out of everyone. Salespeople get commissions, but other people get bonuses that depend on having a Lionel Messi who can hit the goal. It's not enough to have only salespeople playing well. It's a team effort. When we all strengthen each other, the whole company benefits.

If you have questions or need help, I am at your service. *Magical Selling* is the foundation of my "sales-as-a-service" company that targets Series A– and Series B–funded companies that need to scale but don't have the cash to deploy a big sales team. For such companies, we offer enterprise sales training, rent-a-sales teams or manager, rent-a-sales coach, and go-to-market strategies. These are all results-centered, all based on the framework. If you would like to learn more, please visit www.MagicalSelling.com.

I wish you the best as you practice magical selling in the real world, develop mastery, and pass these lessons on to peers, employees, and even management. They need this framework more than anyone.

Thank you for joining me on this journey, but truth be told. . .

Your adventure has only begun.

GO BEYOND THE BOOK.

Ask Raju to help your team win enterprise sales deals like magic at www.MagicalSelling.com.

ABOUT THE AUTHOR

Raju Bhupatiraju drives consistent enterprise IT sales in any economy, across cultures, and at companies of all sizes, from fifteen-employee startups to the Fortune 500. He has a twenty-plus-year track record of achieving quota, turning around underperformance, building partner ecosystems in emerging technologies from scratch, and energizing disparate teams to win big. Raju's specialty is communicating the business outcomes of large, complex, transformational projects. He is a highly sought after Sales-as-a-Service consultant and the author of *Magical Selling: Engineering Enterprise Sales Success on any Team, in any Industry, and in any Economy*. Ask Raju to help your team win enterprise sales deals like magic at www.MagicalSelling.com.

www.ingramcontent.com/pod-product-compliance
Lightning Source LLC
Chambersburg PA
CBHW021924190326
41519CB00009B/902

* 9 7 8 1 7 3 5 8 5 4 6 0 1 *